THE HERBAL OF THE COUNT PALATINE

An eighteenth-century herbal

Tab. IX.

Christoph Jakob Trew

THE HERBAL OF THE COUNT PALATINE

An eighteenth-century herbal

Illustrated by
Elizabeth Blackwell
and
Georg Dionysius Ehret

HARRAP LONDON

Page 2, one of the eighty or so species of the genus Magnolia, *dedicated to Pierre Magnol in the 18th century. Opposite, its fruit and seeds. Left,* a Cereus, *a member of the Cactaceae.*

Illustrations by Elizabeth Blackwell from the *Herbarium Blackwellianum* and by Georg Dionysius Ehret from *Plantae selectae*.

The quotation from Homer's *Odyssey* on page 11 is taken from the translation by E. V. Rieu (Penguin, London, 1946).

The quotations on pages 26, 93, 102 and 104 are taken from Agnes Arber, *Herbals, their origin and evolution* (Cambridge University Press, Cambridge, 1938), by kind permission of the Syndics of Cambridge University Library.

The quotation from the *London Magazine* on page 53 is taken from Marjorie Caygill, *The Story of the British Museum* (British Museum Publications Ltd., London 1981).

The quotations from Casanova's *Mémoires* on pages 61–62 are taken from the translation by Arthur Machen (Elek Books, London, 1959).

The quotation from Plutarch's *On Isis and Osiris* on pages 65–66 is taken from the translation by Frank Cole Babbitt (William Heinemann Ltd., London, 1936).

The quotation from Christopher Columbus's *Journal* on page 93 is taken from the translation by Cecil Jane (Anthony Blond, London, 1958).

Translated by Lucia Woodward.

First published in Great Britain 1985
by Harrap Limited
19–23 Ludgate Hill, London, EC4M 7PD

First published in Italy under the title *L'Erbario del Conte Palatino* by Arnoldo Mondadori S.p.A., Milan.

ISBN 0 245-54302-3 (cased)
ISBN 0 245-54308-2 (paper)

Printed and bound in Italy by Officine Grafiche di Arnoldo Mondadori Editore, Verona.

Christoph Jakob Trew, respected burgher of Nuremberg, avid collector, Count Palatine of the Holy Roman Empire and recipient of many other honours, sponsored, among his many other activities, the publication of two handsome collections of plant illustrations. One was the work of Elizabeth Blackwell, a somewhat ill-fated English lady; the other was painted by Georg Dionysius Ehret, an ambitious man, who, from being a gardener, succeeded in becoming a botanical artist. Together, the two publications represent the broad outlines of a catalogue of the plant world which was being eagerly explored in the 18th century, as part of the wider awakening of interest in nature.

Over one hundred of the species depicted in the two herbals are reproduced in this book. The historical context in which they were painted is evoked by unusual details about the period and by the thoughts, botanical and otherwise, of some contemporary figures. The result is a wealth of vivid insights into the love for plants and flowers which was so typical of the 18th century, when scientific learning flourished in the general climate of discovery and enlightenment.

The physician, the noblewoman and the gardener

Three characters are brought together by this book: Christoph Jakob Trew, Elizabeth Blackwell and Georg Dionysius Ehret. A portrait of the first reveals a somewhat stern-faced man with strongly arched eyebrows, thin lips, and the corners of his mouth slightly upturned as if repressing an ironical smile. The lines running from his nostrils to his cheeks emphasize his prominent cheek-bones; his chin is long and his forehead high. There is an intense, intelligent look in his eyes. The figure depicted in the engraving by J.J. Haid after the portrait by Gabriel Müller (*c.*1768) is Christoph Jakob Trew, Count Palatine of the Holy Roman Empire, court physician, city councillor, and close adviser to the Margrave of Brandenburg-Ansbach.

The book held open in his strong hands is a clue to the physician's intellectual passion: botany. The rococo curlicues of the frame are laden with motifs of fruits, flowers, leaves and foliage. Such motifs were integral to the rococo style, but their exuberance in this case is equivalent to a message. A cartouche bears, in Latin, his professional, civic and academic titles: Director of the Imperial Academy *naturae curiosorum* (of natural curiosities), member of the Royal Society of London, of the Berlin Academy and of the Florentine Botanical Society, among others.

Born in 1695, Dr Trew lived in Nuremberg. The town at the end of the 17th century was very beautiful, wealthy but somewhat sleepy. Within the city walls lay narrow streets of half-timbered burgher houses (like the one in which Dürer had lived, situated on a street climbing the hill to the burgravial castle), large Gothic churches and bridges across the river Pegnitz which cuts the town in two. In the market square was the beautiful 14th-century Schöner Brunnen—the fountain with its forty statues by Heinrich Parler—and the clock of the Frauenkirche, with its daily mechanical procession of seven Electors bowing to Charles IV, in memory of the Imperial Diet held in Nuremberg in the middle of the 14th century when the Golden Bull was issued. The town hall had been recently restored for the twenty-six burgomasters of the Lesser Council and the few hundreds of citizens of the Upper Council who ran the republic. Yet international commerce had ceased to flourish, despite the fact that the town enjoyed a favourable position, situated at the point where the road from the Rhine to the Danube joined that which led from Saxony to the Alpine passes. In Venice, the Fondaco dei Tedeschi, a company run mainly by Nuremberg merchants, was languishing. Financial power had also declined, and with it the eagerness for learning and the artistic activity which had made the town the heart of Humanism and the Renaissance in Germany. The Thirty Years War, which was still remembered by older people, had exhausted the town's resources.

Christoph Jakob Trew lived to the age of seventy-four (1695–1769). He was little more than a child when the War of the Spanish Succession began, and nineteen when it came to an end. Bavaria fought on one side and the Empire, supported by various German princes, on the other. Germany was once again divided in allegiance, and therefore a battlefield itself, during the Seven Years War. Trew died six years after peace was declared. Despite the fact that Germany was divided into some 350 sovereign states (kingdoms, principalities, grand duchies and free towns—Nuremberg being one of the latter), in the period between these wars, it was an industrious, peaceful country which fostered learning and a love of nature and music.

Trew's professional status as a physician is indicated by his position as Head of the Medical Association. He was a learned man with a passion for books—the library he left on his death contained 34,000 volumes—and an academic's interest in botany. His enthusiasm for these two areas found expression in another passion: he loved to sponsor the publication and printing of learned, beautifully illustrated books on plants, at a time when many new botanical species were being discovered.

Two centuries earlier, Konrad Gesner, a Swiss scholar whose encyclopedic knowledge and wide-ranging interests invite parallels with the Roman writer Pliny, had collected an enormous botanical corpus, including some 1,500 drawings of plants, which had never been published. In 1744 Trew bought all this material dating from different periods. The book-loving physician had the satisfaction of seeing about one-third of the drawings he had bought printed in two hand-coloured folio volumes. This was only one of several ventures undertaken by Nuremberg's chief physician. The others are of greater interest in that they involve our two other characters.

Elizabeth Blackwell, the daughter of an Aberdeen hosier, made an unhappy marriage. Her story does not quite possess the tragic quality captured by William Hogarth in his *Marriage à la Mode*, a roughly contemporary series of paintings and prints, but it shares something of its atmosphere. Her husband's life could have been the subject of one of Defoe's novels: Alexander Blackwell was endowed with initiative, a sense of adventure, the desire to live life to the full, an open-mindedness that verged on the unscrupulous and—an unfair share of bad luck. He was apparently of illegitimate but noble birth, or so he liked people to think. He read medicine, first at Cambridge and then at Leiden under the great Dutch naturalist, Herman Boerhaave. He travelled in France, Spain and Portugal, in the tradition of the grand tour undertaken by the English nobility. He practised medicine in Scotland, where he met his wife, and in London. It is not clear why, at one point, he became a proof-reader and later came to own a printing works; what we do know is that he gambled away his wife's dowry and found himself in a debtors' prison.

It was then that Elizabeth came into her own. As a well-brought-up young lady, drawing had very probably formed part of her education, and what more suitable subjects for the wife of a gentleman than plants and flowers? Besides, botany must have played an important part in her husband's work, since medical remedies in those days were based almost entirely on herbs. Elizabeth Blackwell showed some of her drawings to Sir Hans Sloane and other physicians, as well as to members of the Society of Apothecaries, and the idea of publishing a herbal—a collection of plates illustrating various herbs, accompanied by a descriptive text—was born. Elizabeth went to live in Swan Walk which, at that time, was in the country, so that she could pick fresh herbs in Chelsea's botanical garden. She drew some 500 plants and made coloured engravings from the drawings. The plates were first published in weekly instalments, and later in a two-volume collection (1737–39). Whether because the time was right for it, or whether because of the backing she received from Sloane and the Society of Apothecaries, the venture was a success and Elizabeth earned enough money to secure her husband's release.

He, however, did not use his freedom well. Only eight years after the publication of his wife's *Curious Herbal* he was executed. His last years make interesting reading. He took up agronomy and was employed by the Duke of Chandos to improve his land. He also wrote *A new method of improving cold, wet and clayey grounds*. He then moved to

Sweden, not with Elizabeth but with an English merchant and his wife. He seems to have travelled widely in the company of this lady; she was known as his cousin and her husband died suddenly one night of colic, which Blackwell himself treated. Needless to say, gossip was rampant.

In Sweden, Blackwell met with some good luck when he was appointed court physician to King Frederick I, but it was this position which led to his death. Perhaps unintentionally, he found himself involved in a political plot to assassinate the King. The struggle between France and Britain for influence lay behind the plot—the War of the Austrian Succession was raging, the two countries were enemies. Compromising letters and large sums of money were mentioned; Alexander Blackwell was questioned under torture and sentenced to death in 1747. He kept his sense of humour to the end. Having placed his head in the wrong position, he is reported to have said, 'Sorry, but it is my first time—no wonder I need some tuition.'

Elizabeth lived on in London (as far as we know, she never left her homeland) and when she died in 1758 she was buried in Chelsea churchyard. In Nuremberg Dr Trew took an interest in her book. What he had in mind was a *Herbarium Blackwellianum* in five volumes, with Latin text. The plates were to be redrawn and corrected according to the recent teachings of Linnaeus, then engraved by N.F. Eisenberger, formerly court painter to the house of Saxony-Hilburghausen. Trew translated and edited the first volume. After his death, the task was carried on by Christian Gottlieb Ludwig, a professor at Leipzig, and by other eminent friends. A sixth, supplementary volume delayed the publication of Elizabeth Blackwell's work until 1773. Plates from this edition are reproduced in this book, in the chapters *A knowledge of the simples* and *The kitchen garden and the wood*.

Our third character was, like Trew, a German, but he spent over half his life in England. In 1728 the twenty-year-old Georg Dionysius Ehret stopped in Regensburg (formerly Ratisbon), on his way to Vienna, where he hoped to find his fortune. Regensburg became more than a mere staging-post.

Ehret was then no more than a gardener, but one endowed with exceptional talent as a painter. He was born in Heidelberg, and his father, court gardener to the Elector Palatine, taught him how to draw plants. Ehret spent his whole childhood among gardens and gardeners, at the courts of German princes. He was employed in the Heidelberg garden thanks to his step-father, who had succeeded Ehret's father in his professional as well as in his private life. He then moved to the formal gardens at Karlsruhe, which belonged to the Margrave Karl III of Baden-Durlach. The prince took a keen interest in horticulture, particularly in the exotic varieties which were being introduced into Europe, and followed the advances being made in botany at universities in the Netherlands.

While at Karlsruhe, Ehret learnt the basic techniques of watercolour from the court painter, but his desire to improve himself and his position aroused jealousy and hostility among his fellow gardeners and this forced him to leave. At least, this is what he tells us. There is virtually only one account of his life story: his autobiography, which is now in the British Library.

On arriving in Regensburg, Ehret joined the household of the apothecary Johann Wilhelm Weinmann, as one of several apprentices. His contract stated that, in the course of a year, he should produce 1,000 drawings for a botanical work to be published by his master; in return he was to receive board and lodging and fifty thaler. He only completed 500 drawings and was paid thirty thaler. However, he had gained entry to a certain circle and this led to his employment with a banker named Loeschenkohl. Ehret's tasks were to plan the banker's garden and to colour the copper engravings illustrating a formidable Dutch work, the *Hortus indicus malabaricus*.

Ehret spent several years with Loeschenkohl. He completed only a third of his commission but made the most of the opportunity to improve his botanical knowledge, and to enter into correspondence with Dr Trew, whom he subsequently visited in Nuremberg in 1733. Trew was highly appreciative of the painter's talent and their collaboration lasted for the rest of their lives. Trew was probably the only person Ehret did not quarrel with—he always had difficulty in making people accept him as the botanist he considered himself to be, rather than merely as an illustrator or, worse, a gardener. The Regensburg banker was one of those with whom Ehret severed his contacts.

Ehret travelled widely in Switzerland, France, Holland and England. In Basle, a rich gentleman named Samuel Burckhardt employed him to plan his gardens. In Berne, he paid for his lodging with a drawing. When he was on his way to Lyons, a French lady invited him to Paris and later introduced him to the director of the Jardin du Roi.

In Paris, Ehret lodged with the botanist Bernard de Jussieu; he sold his flower paintings to the nobility, sent drawings to his friend Trew and improved his knowledge of botany but, financially, was less well off. In 1735, armed with authoritative letters of introduction, Ehret moved to England. In London he met Sir Hans Sloane, the physician who, in these very years, was encouraging Elizabeth Blackwell in her surprising career. The German painter also profited from the excellent opportunities for botanical observation provided by Chelsea's botanical garden, but he did not settle down immediately and the following year moved to Holland.

In 1736 Linnaeus was working there as physician and botanical adviser to the rich banker George Clifford, and was compiling a scientific description of the species growing in Clifford's private garden near Haarlem. Clifford met Ehret and was so impressed with his work that he commissioned him to draw the plates for Linnaeus' *Hortus Cliffortianus*. However, Ehret's relationship with the Swedish scientist soon deteriorated; once again he did not succeed in being regarded as an equal by botanists, which was his greatest aspiration. Later that year he returned to England, where he spent the rest of his life.

In England, Ehret consolidated his fame as a botanical artist. He worked for the physician Richard Mead, another of Mrs Blackwell's patrons. He lived with the most famous gardener of the time, Philip Miller, and married his sister, only to fall out with Miller later. He was a regular visitor to the house of the Duchess of Portland, a botanical enthusiast, in whose drawing-room he met the most famous botanists of the time. He was offered the directorship of the Oxford University botanical garden by Humphrey Sibthorp, Professor of Botany there, but he left the post shortly afterwards, in 1751, as he found he was unable to reconcile his career as a botanical painter with the practical demands of his job and employer.

As a result of the society he kept—that odd community of physicians, apothecaries, professors, bourgeois and aristocratic gardeners, all fired with the enthusiasm for botany which was so much in vogue—he finally gained recognition in 1757, when he was elected a member of the Royal Society.

Ehret's collaboration with Trew continued with successful and mutually satisfactory results. The chapters of the present volume entitled *The garden of curiosities* and *The plant hunters* reproduce illustrations from the *Plantae selectae*, one of Trew's publishing ventures. The work consisted of a hundred plants, superbly painted by Ehret and engraved by Johann Jakob Haid, which were published in groups of ten plates at a time. Neither the painter, who died in England in 1770, nor the physician, who died in Nuremberg in 1769, was able to see the work completed, as publication began in 1750 and ended in 1775.

The garden of curiosities

In the same enclosure there is a fruitful vineyard . . . Vegetable beds of various kinds are neatly laid out beyond the farthest row and make a smiling patch of never-failing green.

Homer, *Odyssey*, Book VII (The palace of Alcinous)
Translated by E.V. Rieu, 1946

Ribwort plantain

Plantago lanceolata, Plantaginaceae

It is not clear why this plant should have been called veratrum in the past, since its growing habit is very different from that of the various species belonging to the genus *Veratrum*. The only similarity lies in the leaves, which have marked parallel veins. Plantains grow all over the world; the freshly chopped leaves are used in popular medicine to treat bruises.

Ribwort plaintain

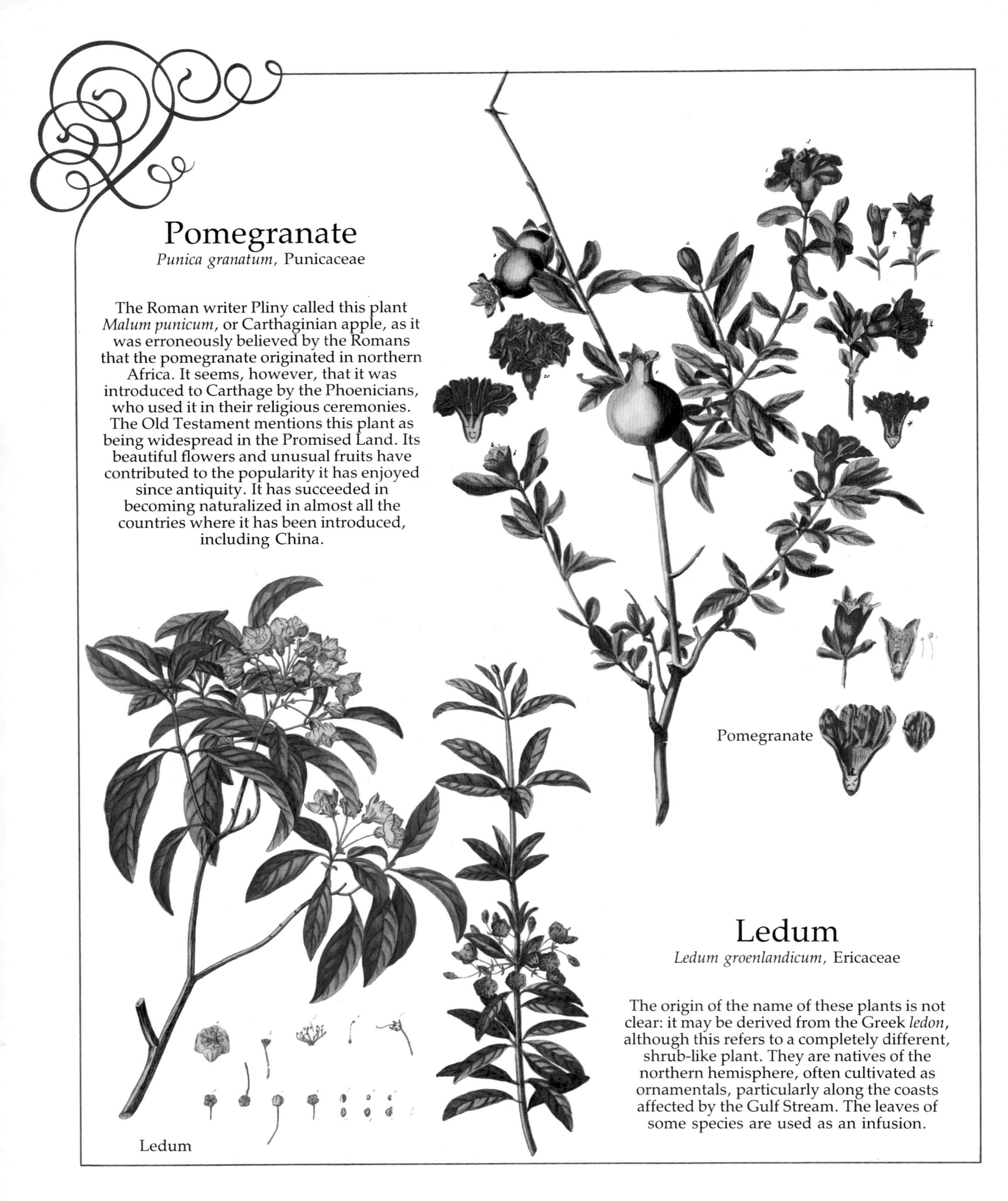

Pomegranate

Punica granatum, Punicaceae

The Roman writer Pliny called this plant *Malum punicum*, or Carthaginian apple, as it was erroneously believed by the Romans that the pomegranate originated in northern Africa. It seems, however, that it was introduced to Carthage by the Phoenicians, who used it in their religious ceremonies. The Old Testament mentions this plant as being widespread in the Promised Land. Its beautiful flowers and unusual fruits have contributed to the popularity it has enjoyed since antiquity. It has succeeded in becoming naturalized in almost all the countries where it has been introduced, including China.

Pomegranate

Ledum

Ledum groenlandicum, Ericaceae

The origin of the name of these plants is not clear: it may be derived from the Greek *ledon*, although this refers to a completely different, shrub-like plant. They are natives of the northern hemisphere, often cultivated as ornamentals, particularly along the coasts affected by the Gulf Stream. The leaves of some species are used as an infusion.

Ledum

Cedar of Lebanon

Cedrus libani, Pinaceae

Cedars are the most important trees in the Lebanon and are widespread in other Mediterranean countries. They have been used and highly prized for their wood since antiquity. Solomon used cedars in building the Temple at Jerusalem and, according to the Greek philosopher Theophrastus, Asian temples used to be clad with cedar wood.

Cedar of Lebanon

Tephrosia

Lungwort

Tephrosia

Tephrosia cinerea, Leguminosae

Some species of tephrosia were once classified as part of the genus *Galega,* as is the case in some old illustrated herbals. They are native to the West Indies, where they are valued for their soporific properties. In particular they are used for fishing in enclosed pools, where their juices make the fish sluggish, and so easier to catch.

Lungwort

Pulmonaria, Boraginaceae

The name of this genus derives from the Latin *pulmo,* meaning lung, because the European species (*Pulmonaria officinalis*) has broad leaves covered in dark spots resembling diseased lungs. In the Middle Ages this resemblance led to the belief that the leaves would cure lung diseases. A characteristic of these plants is the way the corolla changes colour once the flower has been fertilized: this is due to a change in the acidity of the cellular fluids, which alters the pigmentation (anthocyanin) of the corolla from blue to red.

Azalea

Rhododendron, section *Azalea,* Ericaceae

In botanical terms common garden azaleas are classified as rhododendrons, a genus which is subdivided into series of similar species, one of which is named *Azalea*. Greatly admired as ornamental plants, they are widely cultivated both as houseplants and in the open. They thrive in mild, humid climates and on acid soils.

Azalea

Rhododendron

Rhododendron, Ericaceae

A literal translation of 'rhododendron' is 'rose tree', which gives rise to the plant's vernacular name in many languages, such as the German *Alpenrose* (Rose of the Alps). The most beautiful specimens are to be found in the Himalayas, one of the main centres of distribution of the rhododendron. Many of the original Himalayan species have subsequently been crossed to obtain increasingly spectacular varieties.

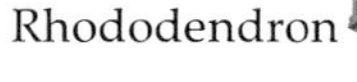

Rhododendron

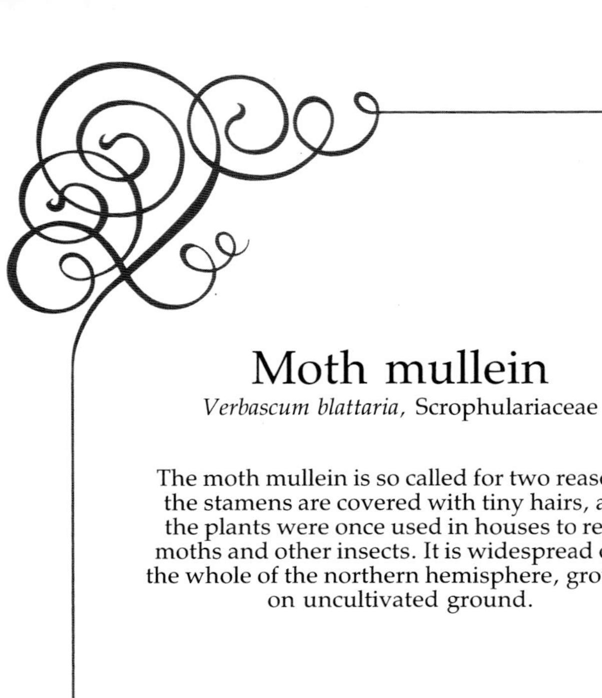

Moth mullein

Verbascum blattaria, Scrophulariaceae

The moth mullein is so called for two reasons: the stamens are covered with tiny hairs, and the plants were once used in houses to repel moths and other insects. It is widespread over the whole of the northern hemisphere, growing on uncultivated ground.

Moth mullein

Iris

Iris

Iris ochroleuca, Iridaceae

The genus *Iris* consists of a large variety of species, which tend to be distinguished by the addition of an adjective describing the plant, for example, stinking or bearded iris, or referring to its place of origin. They are all noted for the beauty of their flowers, many of which are showy and rainbow-coloured. Iris was the name of Juno's messenger, who, on her way to Earth, unfurled her multicoloured shawl, thus creating the rainbow. The plant was mentioned by Theophrastus and was used by the Romans during their religious rites.

Lily

Lilium, Liliaceae

Lilies were renowned for their beauty even in antiquity and appear as decorative motifs on many Etruscan and Greek vases. Pliny, one of history's greatest naturalists, wrote that the lily equals the rose in nobility. So many species and varieties are to be found that this genus can effectively be regarded as being cosmopolitan in distribution.

Lily

Boehmeria

Boehmeria

Boehmeria, Urticaceae

Linnaeus classified the Boehmerias as part of the genus *Urtica*, but in the 19th century they were placed in a genus of their own, named after the German botanist G.R. Boehmer. They are native to the tropics. A group of related species, of which *Boehmeria nivea* is the best known, is widely cultivated for the textile fibre, ramie, which can be extracted from the stems after maceration.

Large red deadnettle

Lamium garganicum, Labiatae

Many species of this genus are widespread in Europe and are often mistaken for nettles because of the shape of their leaves. The flowers are throat-shaped.

Large red deadnettle

Fig

Ficus carica, Moraceae

This species is widely cultivated in the Mediterranean area and is well known for its delicious fruits. They are actually false fruits, in that they are the seed receptacles.

Fig

Periploca

Periploca, Axclepiadaceae

The 16th-century Sienese physician Pierandrea Mattioli explains in his *Discorsi* how these plants got their name and how they came to be classified in a different genus from their original one, the *Apocynum.* He noticed that from the seeds of two *Apocynum* plants, one grew trailing or climbing stems (hence periploca, meaning voluble, winding). In the past the juice of these plants was mixed with bait and used as an effective poison for rats and mice.

Periploca

Sassafras

Sassafras

Sassafras variifolium, Lauraceae

The generic name of this tree is derived from the Spanish *salsafras,* which designates certain saxifrages, because when the sassafras was discovered it was believed that it had medicinal properties similar to those of the saxifrages. The foliage of the sassafras is unusual in that some of the leaves are entire and some trilobate. They turn a beautiful yellow, then red, in the autumn, so the tree is often planted as an ornamental.

The garden of curiosities

The first botanical gardens

Botanical gardens have been as important in the development of botany as anatomy class-rooms in the development of medicine. With the spirit of enquiry and curiosity which characterized the 18th century, such gardens were even more popular and in vogue than they are today. The Chelsea botanical garden was probably the best known until Kew became pre-eminent. It was not, however, the first to be established.

Botanical gardens originated in Italy in the first half of the 16th century. What little botanical knowledge existed at that time was entirely subordinated to medicine. Herbalists and apothecaries who, since classical times, had prepared and sold drugs, supplied what were known as simples (from the Latin *medicamentum simplex*, or simple drug), that is, plants with medicinal properties and ingredients necessary for the preparation of medicines. Indeed, it was a Professor of Medicine at Padua University, Francesco Bonafede, who first suggested to the Venetian Senate that a chair of *lectura simplicium* (instruction on the simples) should be created. This was done in 1533 and the chair went to Bonafede himself, who thus became the first university professor in botany.

The following year, Padua's example was followed by the University of Bologna, where Luca Ghini was nominated *lector simplicium*. Several years later, in 1561, a second chair was created at Padua, designated *ostensor simplicium in horto*, the aim of which was to show students the medicinal plants in their natural surroundings and to teach them how to recognize them. With the introduction of this practical approach, botany ceased to rely solely upon learned texts. In the meantime, the first botanical gardens had been created. They were founded in Padua and Pisa by Bonafede and Ghini respectively, between 1543 and 1545 and were immediately followed by one in Florence.

At Padua the garden is circular, built in the shade of the domes of the Chiesa del Santo, the church dedicated to the city's patron saint, St Anthony. The garden is divided into geometrical beds according to the canons of Renaissance design. The great German poet Goethe visited the garden in 1786 and a palm tree he saw there inspired his theory of metamorphoses. Goethe's palm is still there today. It had been brought from Egypt at the end of the 16th century by Prospero Alpino, another holder of this oldest Chair of Botany.

The Pisan botanical garden introduced the camellia and the horse-chestnut to Italy. Luca Ghini succeeded in convincing the Medici duke, Cosimo I, to found the garden. Ghini, the son of a notary from Imola, was a physician who devoted his whole life to botany. As we have seen, he was Professor at Bologna; in 1544 Cosimo summoned him to Pisa. Ghini used to travel around Italy with friends and students, gathering plants. He was also sent samples from various correspondents in Egypt, Syria, Spain, Sicily and Calabria, and from his brother who practised law in Candia (Herakleion).

In a letter dated 4 July 1545, probably addressed to an official at the Medici court, Ghini writes: 'I have already paid two visits to the mountains and gathered many beautiful plants which I have had planted with the greatest ease in a large garden in Pisa. However, I fear that this heat will make them wither and therefore would implore your Grace to be so kind as to prepare the garden at Pisa, as I wish to create a garden that will please his Excellency and be useful to the students.' The Duke himself had a passion for medicinal botany and, as his biographer Baccio Baldini wrote, he had all sorts of herbs, leaves and flowers distilled throughout the year, thus obtaining 'waters and highly precious oils'.

Again prompted by Ghini, the Duke founded the botanical garden in Florence, probably at the beginning of 1546. He called in the sculptor Niccolò Pericoli, known as Il Tribolo, who later worked on the Boboli Gardens. Il Tribolo drew a rectangular plan divided into four by two avenues; each section was in turn divided into beds of varying shapes, consisting of an intricate pattern of circles, squares and leaf motifs. A special aqueduct was built to bring water from a reservoir on the river Mugnone.

The development of the herbal

The examples of Padua and Bologna were soon followed by the other universities, and botanical gardens were copied elsewhere in Italy and abroad. The appointment of professors to teach botany and the creation of botanical gardens are both phenomena which should be regarded as part of the more general cultural awakening of the Renaissance. They brought about an important development in the traditional herbal. Throughout the Middle Ages the Latin word *herbarius* indicated a treatise, a manuscript catalogue of plants with or without illustrations. In the 16th century the herbal began to be supplemented by dried plants mounted on sheets of paper and was called a *herbarium* or occasionally *hortus hiemalis* (the winter garden) or *hortus siccus* (the dry garden). However, the recent development of printing meant that written herbals, far from being superseded, came to enjoy unprecedented popularity, which had important implications for both art and science.

The greater availability of both types of herbal was one of the factors that led to the triumph of naturalism in painting over the stylized, stock images of medieval art. A tendency towards naturalism is already evident in Botticelli's *Spring*—of the 190 flowering plants depicted, 138 have been identified by botanists. It gained impetus when artists were able to copy plants from specimens in the *hortus hiemalis* and in the botanical gardens, as Elizabeth Blackwell and Georg Dionysius Ehret were to do.

It has been suggested that the *hortus siccus* may have been invented by Ghini, but if this is so, his own herbal has not survived. We know that he sent dried plants glued on to sheets of paper to the Sienese physician Pierandrea Mattioli, commentator on the Greek author Dioscorides, and that Ghini had a collection of 300 of these specimens. We do have herbals belonging to his pupils: that of Fr Michele Merini contains 202 plants, mainly from the Tuscan coast; another in five volumes, portraying 1,600 plants, is in the Biblioteca Angelica, Rome. It is dated 1532, and may be the oldest existing herbal. It was probably compiled by Gherardo Cibo, who attended Ghini's lectures at

Bologna. Luca Ghini's example was certainly responsible for dried herbals being commonly used among botanists, and this led to an eager exchange of specimens from one country to another.

A considerable amount of research has also been devoted to discovering when plants began to be painted directly from nature with the aim of producing faithful illustrations for written herbals. It was once thought that the earliest example was the *Liber de Simplicibus* (Book of Simples), a Venetian codex dated 1419, in the Biblioteca Marciana, Venice, by Benedetto Rinio and illustrated by Andrea Amadio. But, as Wilfrid Blunt argues in his book *The Illustrated Herbal*, Rinio's herbal is not quite the oldest since classical times to contain illustrations painted directly from nature, for about twenty of his plant paintings were copied from a slightly older codex, now in the British Library in London.

This brings us back once more to Padua, the city that played such an important part in the history of botany. The codex in question is the *Herbolario volgare* (Popular herbal), an Italian translation of a treatise on botanical pharmacology written by the 9th-century Arab physician Serapion the Younger. The Italian text was written at the end of the 14th century by the Paduan monk Jacopo Filippo, who left blank spaces for the illustrations. Only about fifty were ever painted, all in the exquisitely elegant style of northern Italian Gothic. The work was commissioned by Francesco da Carrara the Younger, the last lord of Padua.

Impressions of a visitor to England

'It is a great disadvantage not to be a botanist when one is travelling on foot.' Thus wrote Giuseppe Baretti, a man of letters from Turin who lived for a long time in England. He published a series of *Lettere familiari a' suoi fratelli* (Letters to his brothers) written while he was travelling back to Italy via Portugal, Spain and France. Referring to two contemporary botanists with whom he was acquainted, he continued, 'Doctors Alione and Marsili, from Padua, must have envied me the good luck which allowed me to wander at will in the area of the Venta do Duque (Portugal), and I would gladly have granted them the privilege of doing so in exchange for this pinch of tobacco which I am holding in my left hand, while with my right I continue my story. When Dr Marsili was staying with me in London, we sometimes strolled together in the botanic gardens in Chelsea, and I would ask him the name of this and that plant – names which I forgot within a few minutes' (20 September 1760). Chelsea at that time was a village about a mile from London.

Botanic gardens at Chelsea and Kew

The botanical gardens in Chelsea are now called the Chelsea Physic Garden. Like its predecessors, this garden was established to further the therapeutic use of plants. It used to be the garden of the Society of Apothecaries and was founded in 1673; eleven years later a conservatory was added to it, heated by a stove installed in the basement. Heated glass-houses such as this were quite a novelty—it appears that the first was used in Heidelberg in 1619 in order to grow orange trees.

John Evelyn, a member of the Royal Society who visited the Chelsea gardens in 1685, was able to admire 'a collection of innumerable rarities', including 'the tree bearing the Jesuit's bark which had done such wonders in quartan' (the bark of trees of the genus *Cinchona* from which quinine is extracted). Another person closely linked with the Chelsea botanical garden is Sir Hans Sloane—the physician, naturalist and collector—who studied there as a student, bought it and later left it in perpetuity to the Society of Apothecaries.

In 1772, the Perfector Horti—the director of the garden—had some forty tons of old stone from the Tower of London arranged in the garden, together with lava transported from Iceland by Sir Joseph Banks, the naturalist who accompanied Cook on the first expedition around the world. The gesture was perhaps influenced by the prevailing taste for rocaille (ornamental rockery and shell forms); it was typical of the cultural atmosphere of the 18th century in which scientific curiosity, love of nature and an aestheticism dominated by the exotic and the fantastic were inextricably linked.

It was this cultural climate which gave rise to the popularity of botanical illustrations such as those which are reproduced here. Another visitor to the botanical garden was Georg Dionysius Ehret, who described it as the *hortus curiosorum* or 'the garden of curiosities' in the epigraph introducing the ten volumes of *Plantae selectae* published by Trew between 1750 and 1775. Each volume contained ten plates drawn by the masterly hand of Ehret from specimens growing in the Chelsea gardens (*Londinii in hortis curiosorum nutrita*). The drawings were engraved on copper, and then delightfully coloured by Johann Jakob Haid—a selection has been reproduced in this chapter.

At about the same time the Scottish botanist and gardener William Aiton was working at Chelsea. He was regarded as the greatest expert on tropical and subtropical plants in Great Britain. A portrait shows him holding a sprig of an evergreen from South Africa which Linnaeus named *Aitonia capensis* in his honour. In 1759 Aiton moved from Chelsea to Kew Gardens. In a sense, this marked the beginning of Chelsea's decline, since all the exotic plants that were being discovered were sent to Kew from then on. Kew was also favoured by Princess Augusta, widow of Frederick, Prince of Wales, and mother to Prince George who, in 1760, became George III.

The history of Kew Gardens provides interesting insight into the life of the aristocracy and the cultural concerns of the 18th century. At that time the huge area of approximately 300 acres covered by the present-day gardens on the south bank of the Thames was divided into two separate properties by Love Lane, a bridle-path between Richmond and Brentford ferry. In 1730 the western part overlooking the river—the gardens of Richmond Lodge, which once stood in the Old Deer Park—became the residence of Queen Caroline, wife of George II, who was described unflatteringly by Horace Walpole as a woman who made great pretensions to learning and taste with not much of the former and none of the latter. Her son, Prince Frederick, whom she regarded as 'the greatest ass, and the greatest liar, and the greatest *canaille*, and the greatest beast, in the whole world', went to live in the other half of the property.

Queen Caroline had a Hermitage built on her land—a mound covered by shrubs and crowned by three conifers, with a façade built of crumbling stone. Inside was a room with niches containing the busts of such illustrious personages as Robert Boyle, John Locke and Sir Isaac Newton. Later, the Queen had the Merlin Cave built, probably by the architect William Kent. It consisted of three rooms with conical thatched roofs and a Tudor portal; the pavilion was approached through several fine walks and agreeable labyrinths. Inside, a group of wax figures was seated round a table; among them were Queen Elizabeth I and the Queen of the Amazons, all in consultation with Merlin. Another room housed a library in the charge of the poet Stephen Duck who, from being a farm labourer became the Queen's protégé, and was consequently a fashionable visitor for a brief period in the drawing-rooms of the wealthy, although he then had to settle for a modest ecclesiastical career. Both the Hermitage and the Merlin Cave were later demolished when the gardens were redesigned by Capability Brown, but they were recorded in contemporary prints. An extraordinary expression of the age in which they were built, they perhaps confirm Walpole's opinion of Queen Caroline.

In 1736, at the age of twenty-nine, Prince Frederick married the seventeen-year-old Princess Augusta of Saxe-Gotha, who was charming although not beautiful and, like many people in those days, scarred by smallpox. They shared a passion for gardening; it seems that the Prince in fact died as a result of it, for, while lingering outside in the wet to watch some trees being planted, he caught a cold which turned into pneumonia. During her long widowhood (1751–72) Augusta indulged her passion for gardening with the help of three men: the gardener William Aiton, the architect Sir William Chambers and Lord Bute, an amateur botanist (who was also to die in the pursuit of botany in 1792, as the result of a fall while trying to reach a rare plant).

We have already met Aiton. He was in charge of the small section of Kew, covering only four acres, which was to be the exotic garden. In 1789 he published his *Hortus Kewensis*, a work in three volumes describing 5,600 plants. Chambers was born in Sweden in 1723 and in his youth sailed on the ships of the Swedish East India Company, travelling as far as Canton. He built more follies on Princess Augusta's land than had been erected on Queen Caroline's: a Chinese pagoda (now Kew's emblem); a temple of the Sun which was based on the circular late Roman temple at Baalbek and was destroyed in 1916 when a huge cedar of Lebanon fell on it during a storm; a temple of Aeolus; a ruined arch; a mosque and a house of Confucius, in the style of Chinese architecture. Some twenty buildings were completed of which only six have survived. They included an observatory which was later transferred to the Old Deer Park to allow George III to view the transit of Venus over the sun on 3 June 1769—the event which Captain Cook went to Tahiti to see.

John Stuart, 3rd Earl of Bute, had met Prince Frederick by chance in 1747, one day when rain interrupted the Egham races and a fourth player was needed for a game of whist; the Prince liked him immediately and drew him into his own circle. After Frederick's death, Augusta placed Bute completely, although unofficially, in charge of the gardens at Kew. It was rumoured that the two were also lovers, and the presumed affair was the subject of libellous pamphlets and cartoons.

The young Prince George developed a deep friendship and great admiration for Bute, who was his tutor, so much so that in 1762, two years after his grandfather's death and his own accession to the throne as George III, he made Bute his Prime Minister. This proved to be Bute's undoing. It was said at the time that he was unsuited to being Prime Minister because he was a Scot, the King's favourite and 'an honest man'. Within a year Bute resigned. According to historians such as Trevelyan, the methods used by Bute to secure the Treaty of Paris in 1763 led to the loss of the admiration and prestige Britain had gained abroad during the Seven Years War. The Earl fell from royal favour in 1765, after the King's first attack of porphyria. He retired from politics to live with his wife on his island at the mouth of the Firth of Clyde. During this period, Linnaeus named the *Stuartia malacodendron* after him, a plant which had recently been introduced from Carolina. His pursuit of botany was given great scope when his wife inherited over a million pounds from her father in 1771.

George III transformed Kew Gardens into a single property, joining the half which had belonged to his grandmother, Queen Caroline, and that belonging to his mother. The leading light at Kew during the last years of the 18th century and at the beginning of the 19th was Sir Joseph Banks; while he was there (he was made Director of the gardens), almost 7,000 new species were introduced into England. Botanical knowledge was making great strides.

Strange creatures in early herbals

According to Dr Johnson, travelling counterbalances imagination with reality: instead of dreaming about how things might be, it teaches one to observe them as they are. Dr Johnson was a contemporary of Christoph Jakob Trew, Elizabeth Blackwell and Georg Dionysius Ehret. The Literary Club of which he was a founder member was frequented by such figures as the actor David Garrick, the orator and politician Charles James Fox (who made Lord Bute's life as Prime Minister rather difficult) and Giuseppe Baretti, the Piedmontese man of letters whom we quoted above. Johnson's fundamental good sense was indicative of the spirit of the age.

The empirical observation of reality he refers to was much needed in the field of botany. Medieval botanical texts are fascinating, with their plethora of information and illustrations. However, the impression gained from examining them, or from reading modern histories of botany, is one of difficulty in separating what were valid observations of the natural world from the obfuscation of bookish wisdom which relied upon the authority of classical authors, or from systematic plagiarism interspersed with fantastic, inaccurate details. The outlandish nature of the latter is evident in the following examples from pre-18th-century herbals.

In an Italian text of 1585, the *Herbario Nuovo* (New Herbal) by Castore Durante, physician to Pope Sixtus V, there is an illustration of a strange tree with a few, large, toothed leaves drawn against a background of the moon and stars. The lower part of the trunk is shaped like a woman's body. This tree was called *Arbor malenconico* or *Arbor tristis* (sad tree); its flowers remained closed during the day and only opened at night, giving off an exquisite scent. It was a tree which grew in India from the ashes of the beautiful daughter of a powerful lord (a certain Parisataccho); the girl committed suicide because the sun did not return her love. That was why at dawn, as soon as the tree which had blossomed in the moonlight was touched with the sun's first rays, it withered and appeared to be dead.

John Gerard was a lad from Cheshire, who crossed the Baltic into Russia and travelled from Narva to Moscow before becoming the director of Lord Burleigh's gardens in London. In 1597 he wrote *The Herball or Generall Historie of Plantes*; it contains what may be the first illustration of the potato (the frontispiece shows the author holding a flowering sprig of the potato plant), but it also describes the *Britannica concha anatifera*, or shell tree, from which geese were said to hatch in the north of Scotland and in the Orkneys. Although the author assures us 'but what our eies have seene, and hands have touched, we shall declare', he continues by describing how, on an island off the Lancashire coast, the timbers of wrecked ships have been known to produce first a sort of froth, then shells from which birds' legs emerge, followed by a creature which slowly matures and then drops into the sea, grows feathers and becomes 'a foule, bigger then a Mallard, and lesser than a Goose'.

This was an ancient legend, discredited by Albertus Magnus as early as the mid-13th century, but traces of it still survive in the English language: a goose barnacle (*Lepas fascicularis*) is a crustacean which clings to the planks of ships and has a flat shell with stalk-like growths; a barnacle goose is the wild goose (*Branta* or *Anas leucopsis*) which breeds in the Arctic seas among driftwood.

Another extraordinary specimen is the vegetal lamb, depicted in the successful work by John Parkinson, herbarist to Charles I, *Paradisi in Sole Paradisus Terrestris. A garden of all sorts of pleasant flowers which our English ayre will permitt to be noursed up* (1629). The frontispiece illustrates the Garden of Eden. Amid palms, leafy trees, daisies, sunflowers, tulips and prickly pears, Eve is running towards an open flower while Adam has just picked the apple. The vegetal lamb is impaled on a short tree trunk. It was said to grow in Tartary, in the area of Samarkand. Parkinson's herbal, in keeping with a long tradition, contains much practical advice on the use of herbs; for instance, the juice of the catmint (*Nepeta cataria*) when drunk with wine was supposed to help in treating bruises incurred by a fall or some other accident.

Botany flourishes in the Low Countries

Gerard's herbal owed much to the work of two Flemish botanists, Dodonaeus and Lobelius. After the developments in botany in Italy in the 16th century, the centre of activity moved to the Netherlands (see A.G. Morton, *History of Botanical Science*, 1981). The prosperity brought by the expansion of commerce, together with the development of towns and agriculture, encouraged lavish spending on the creation of pleasant gardens. The work of numerous botanists there, of whom the most brilliant was Clusius, laid the foundations for the expansion of botanical knowledge and eventually led to the replacement of the herbal by the flora as the standard work. At the end of the 15th century herbals covered no more than 1,000 plants; derived from classical and medieval sources, they represented the extent of botanical knowledge. Little more than a century later, 6,000 types of plants were known.

Rembert Dodoens (Dodonaeus) was born in 1517 at Malines, where he practised medicine. He was appointed chief physician to the Viennese court of Maximilian II and Rudolf II, and then taught medicine at Leiden. Charles de l'Ecluse (Clusius) was born in 1525 at Arras and studied medicine, although he never practised it. He too was summoned to Vienna by Maximilian II, and died while teaching at Leiden. While he was Keeper of the Imperial Garden in Vienna, he imported tulip seeds and bulbs from Turkey in 1573, cultivated them and sent them to the Low Countries. He could therefore be considered the father of the Dutch bulb industry; a friend of his, the Princesse de Chimay, called him 'the father of all the beautiful gardens of this country'.

Matthias de Lobel (Lobelius) was born in Lille. He was also a physician, lived in England during the reign of Elizabeth I, became physician to William the Silent in Holland and finally returned to England to be botanist and physician to James I. Although they spent much of their lives abroad, these three Flemish scientists were close friends and collaborators, and all three had works published by the same publisher, Plantin of Antwerp.

Christophe Plantin was a native of Touraine; in the middle of the 16th century he settled in Antwerp as a bookbinder and leather craftsman. Flanders at this time was under Spanish rule. It so happened that the secretary to Philip II had to send a precious jewel post haste to the Spanish Queen and commissioned Plantin to produce a suitable casket. Having finished it late in the evening, Plantin set out to deliver it. On the way he met a party of drunken revellers who had a grudge to settle with a musician. They mistook Plantin for the latter and as a result of being wounded in the sword-fight which ensued, Plantin found the physical demands of bookbinding too much and turned to printing. His house in Antwerp market square is now a museum dedicated to the publishing firm of Plantin-Moretus, founded in 1576. Moretus married one of Plantin's daughters and carried on the business, which survived for nearly 300 years. A sonnet written by Plantin, the first four lines of which are given below, expresses this wise and hard-working craftsman's ideal of a life characterized by a serene independence:

Avoir une maison commode, propre et belle,
Un jardin tapissé d'espaliers odorans,
Des fruits, d'excellent vin, peu de train, peu d'enfants,
Posseder seul, sans bruit, une femme fidèle.

(To have a comfortable, clean and beautiful house, a garden lined with scented espaliers, to have fruit, excellent wine, little excitement, few children, to possess in solitude and tranquillity a faithful wife.)

A botanical fable

'There was a garden in which carnations and delicate pink roses reigned supreme over all the other flowers. They looked down upon the little violets which hid among the grass so that they could barely be seen. The carnations said: "We are so gaily and so variously coloured that all the men and women who stroll in this garden look at us and never cease to admire us." "And as for us," said the roses, "why, all young maidens admire us and carefully pick us to adorn their breast. When our petals are crushed they exude a liquor which fills all the air around with a marvellous scent. We do not know what the violet could possibly have to say for itself, since its scent can barely be detected and its colour is neither as bright nor as showy as ours." The sweet violet replied thus: "Oh most noble among all flowers, nature confers qualities upon us all. You were made to provide a more striking and wondrous ornament in human eyes, and I to adorn the short grass and to add grace and variety to this expanse of green that surrounds me. Everything is good in nature. Some things are easier to admire, but that is no reason to despise the lesser ones."

'The moral of this story concerns virtue. Some virtues are great and noble: magnanimity, for instance, and clemency, and others so outstanding that everyone admires and praises them. But they are not always called for, neither is there opportunity for everyone to put them into practice. In contrast, meekness, humility and affability are within anybody's reach, and although they are neither obvious nor great like the aforementioned virtues, they enrich our everyday life. Indeed, they may do more to bring beauty into the world than the others because they are brought into play by almost all our actions. The former are worthy of a place in history, the latter of being loved by everybody.'

This story appeared in 1761 in *L'Osservatore veneto* (The Veneto Observer), a periodical of selected prose published painstakingly by Count Gasparo Gozzi in Venice and, indeed, written almost entirely by himself.

A knowledge of the simples

I must up-fill this osier cage of ours
With baleful weeds, and precious-juiced flowers.
The earth that's nature's mother, is her tomb;
What is her burying grave, that is her womb.
And from her womb children of divers kind
We sucking on her natural bosom find;
Many for many virtues excellent,
None but for some, and yet all different.
O, mickle is the powerful grace that lies
In herbs, plants, stones, and their true
qualities . . .

W. Shakespeare, *Romeo and Juliet*, II.iii.7–16

Opium poppy

Papaver somniferum, Papaveraceae

Known since antiquity, the opium poppy became particularly widespread in the Middle East when the Arabs took to eating opium after Muhammad forbade them alcohol. Only later did opium extracted from the unripe fruits come to be smoked. At first, the Portuguese had an almost total monopoly of opium distribution, but after their defeat in 1773 at the hands of the British, the opium trade was carried on by the East India Company. Because of the harmful effects of the drug China prohibited it, thereby interfering with the financial interests of the Company. This led to the Opium Wars; China was twice defeated and in 1858 was forced to allow opium to be imported and sold.

Opium poppy

Dwarf mallow

Malva neglecta, Malvaceae

Mallows were known in classical times for their emollient properties; indeed their name is thought to derive from the Greek *malakós*, meaning soft. The famous 16th-century Sienese physician Mattioli recommended the application of mallow, honey and salt poultices to relieve bee stings, and maintained that they could be avoided altogether by covering one's body with a paste of mallow crushed with oil. The generic name *Malva* was the suggestion of the French botanist and court physician Tournefort.

Dwarf mallow

Hollyhock

Hollyhock

Alcea rosea, Malvaceae

This is purely a cultivated ornamental which is never found growing wild or as a naturalized species. Like the mallow, it is valued for its emollient properties. Theophrastus, a contemporary of Aristotle regarded by Linnaeus as the 'father of botany', maintained that the hollyhock takes on a tree-like habit when grown in gardens and, in the same vein, Mattioli wrote that he had seen some beautiful, large, tree-like hollyhocks growing in the gardens of a villa in Grignano on Lake Garda.

Rose

Rosa, Rosaceae

The genus *Rosa* is easily recognizable but the classification of specimens into species is difficult because wild hybrids are very common. There are also endless cultivated varieties, some of which are extremely valuable. For example, 'William Francis Bennet' (varieties are usually named after the person who created the hybrid, or are dedicated to someone) was sold for $5,000.

Rose

Honeysuckle

Honeysuckle

Lonicera caprifolium, Caprifoliaceae

The honeysuckle grows wild all over Europe and is widely cultivated to cover fences, arches and pergolas; it produces masses of heavily scented flowers with long, tubular corollas. The flowers are pollinated by a butterfly which hovers in front of the opening, in much the same way as the humming-bird does with orchids; the insect then inserts its long proboscis in the tubular corolla to reach the nectar at the bottom.

Clove

Eugenia caryophyllata, Myrtaceae

The cloves used in cooking are the aromatic flower buds of this plant. The Dutch destroyed the trees they found growing wild in the Moluccas, which had belonged to Portugal, and monopolized their cultivation on Java and Sumatra. However, in 1770 the Frenchman Poivre found a few plants still growing on the Moluccas and introduced them to Martinique.

Clove

Mezereon

Mezereon

Daphne mezereum, Thymelaeaceae

In Greek mythology, Daphne was the beautiful daughter of Peneius, a river deity, who, in order to save her from the amorous intentions of the god Apollo, changed her into a shrub with leaves similar to those of the bay (*daphne* in Greek). The strongly scented flowers appear in late winter on bare branches. The plant is common throughout Europe and, although it is poisonous, in the past its seeds were used, rather dangerously, to adulterate black pepper.

Thorn apple

Datura stramonium, Solanaceae

A native of tropical America, the thorn apple was introduced into Europe at the beginning of the 17th century as an ornamental. It has since become naturalized and can be found on waste land or sandy littorals. It is poisonous but is used in cases of hay fever.

Thorn apple

Spring adonis

Adonis vernalis, Ranunculaceae

Native to the European and Siberian steppes and naturalized in certain areas of the Alps, the spring adonis was valued as a medicinal plant in the Middle Ages and it is still used for its cardiotonic properties. According to myth, Aphrodite changed her beloved Adonis, the son of King Cinyras, into this flower when the youth was about to die after being gored by a wild boar.

Spring adonis

Lily-of-the-valley

Lily-of-the-valley

Convallaria majalis, Liliaceae

This highly scented flower is a symbol of purity, humility and grace; it is widespread throughout Europe. According to an English legend, this flower grew from the drops of blood shed by St Leonard during his three-day-long fight against the dragon Sin. Linnaeus named it *Lilium convallium*. Both *convallium* and *convallaria* echo the Latin *convallis*, meaning a narrow valley between steep hills.

Stemless thistle

Carlina acaulis, Compositae

This thorny, stemless thistle was named after Charlemagne who is said to have experimented with its medicinal properties to cure himself and his soldiers. It is quite common in Europe and is similar in taste to the globe artichoke.

Stemless thistle

Mandrake

Mandragora officinarum, Solanaceae

A small, herbaceous perennial flowering in spring, the mandrake's thick root vaguely resembles the human body, and ancient herbals distinguished between two species, male and female. It is native to the Mediterranean but is not common. Mandrake roots have been found in the royal tombs at Thebes. According to an Egyptian legend, Ra, the sun god, sent Mathor to earth to punish mankind who had displeased him. Mathor slaughtered so many men that Ra took pity on the human race and forced Mathor to drink the blood of his victims mixed with mandrake root. Drugged, Mathor fell asleep and when he woke up he had forgotten why he had been sent to earth, thus putting an end to the slaughter.

Mandrake

Marsh mallow

Marsh mallow

Althaea officinalis, Malvaceae

A medicinal plant used for thousands of years; the marsh mallow's medicinal properties were acknowledged by both botanical classifiers Tournefort and Linnaeus (the Greek verb *althomai* means to cure). The whole plant is rich in mucilages which give it its emollient, refreshing and soothing qualities. In the past the properties attributed to it were far more numerous: it was used in cases of kidney stones, gonorrhoea, toothache, vesical calculus, St Anthony's Fire, constipation, coughs, asthma and many other complaints.

Christmas rose

Helleborus niger, Ranunculaceae

The Christmas rose flowers even on snow-covered ground. Legend has it that an angel made it appear before a young shepherdess who was sad because she was not able to join the Magi in taking presents to the Christ Child. It was once used to treat depression and epilepsy.

Christmas rose

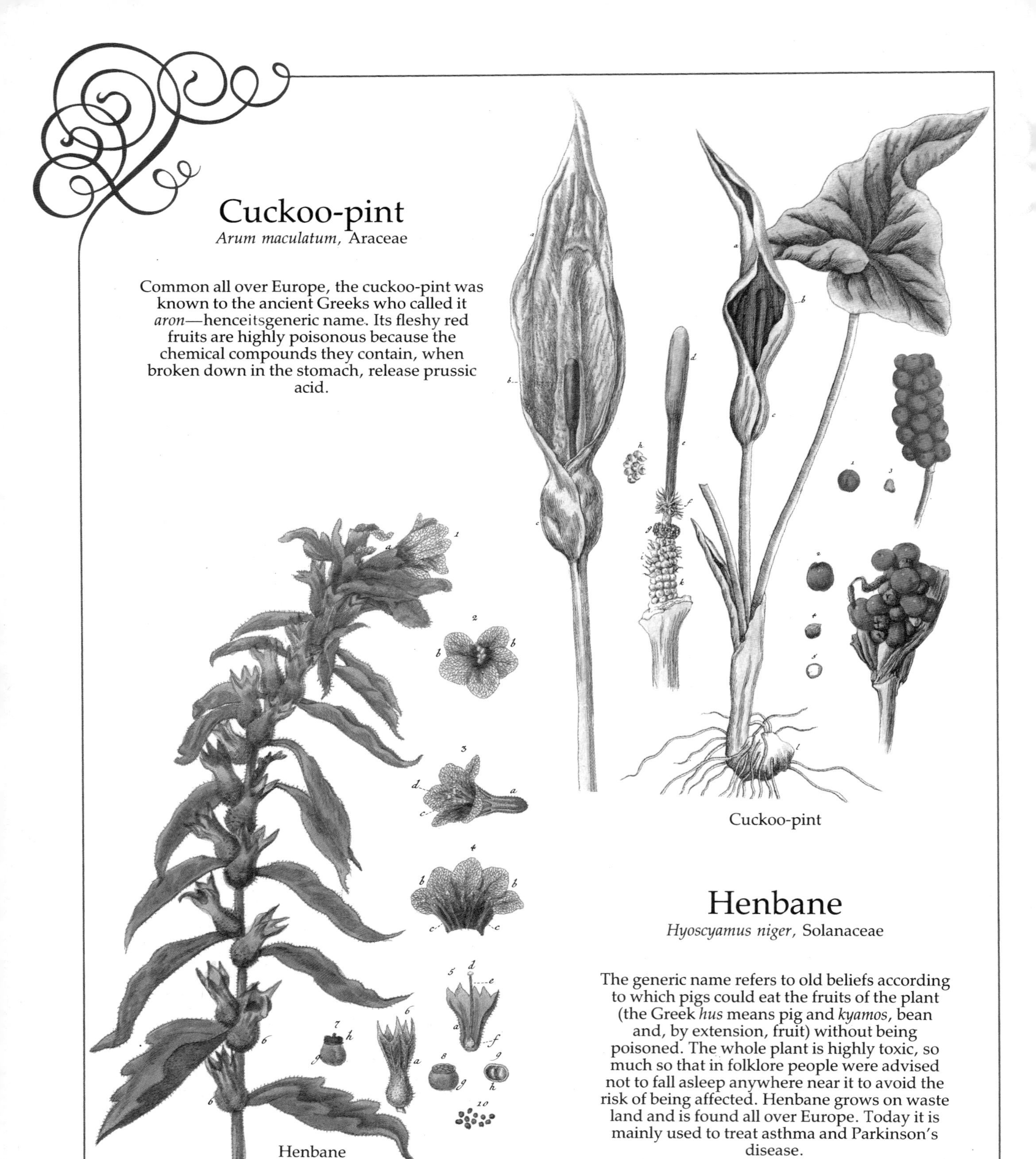

Cuckoo-pint

Arum maculatum, Araceae

Common all over Europe, the cuckoo-pint was known to the ancient Greeks who called it *aron*—henceitsgeneric name. Its fleshy red fruits are highly poisonous because the chemical compounds they contain, when broken down in the stomach, release prussic acid.

Cuckoo-pint

Henbane

Hyoscyamus niger, Solanaceae

The generic name refers to old beliefs according to which pigs could eat the fruits of the plant (the Greek *hus* means pig and *kyamos*, bean and, by extension, fruit) without being poisoned. The whole plant is highly toxic, so much so that in folklore people were advised not to fall asleep anywhere near it to avoid the risk of being affected. Henbane grows on waste land and is found all over Europe. Today it is mainly used to treat asthma and Parkinson's disease.

Henbane

Rosemary

Rosmarinus officinalis, Labiatae

The etymological meaning of *Rosmarinus* is 'sea dew'; quite often, in damp weather, and in the morning, its leaves appear to be covered with tiny droplets of water. Rosemary is common throughout Europe. The Greeks, Romans and Arabs wove its branches into triumphal garlands, as they did laurel leaves. It was one of the seventy-three useful species listed by Charlemagne and is now widely used both in medicine and as a culinary herb.

Rosemary

Liquorice

Liquorice

Glycyrrhiza glabra, Leguminosae

The generic name *Glycyrrhiza* means plant with sweet roots. Today, liquorice grows wild in southern and central Europe and Asia and is cultivated elsewhere. It was known in antiquity and was used in the sacred rite of Buddha's bath. It appears that Benedictine monks cultivated it in Italy and introduced it into Spain in the 10th century, thus spreading the knowledge of its medicinal properties.

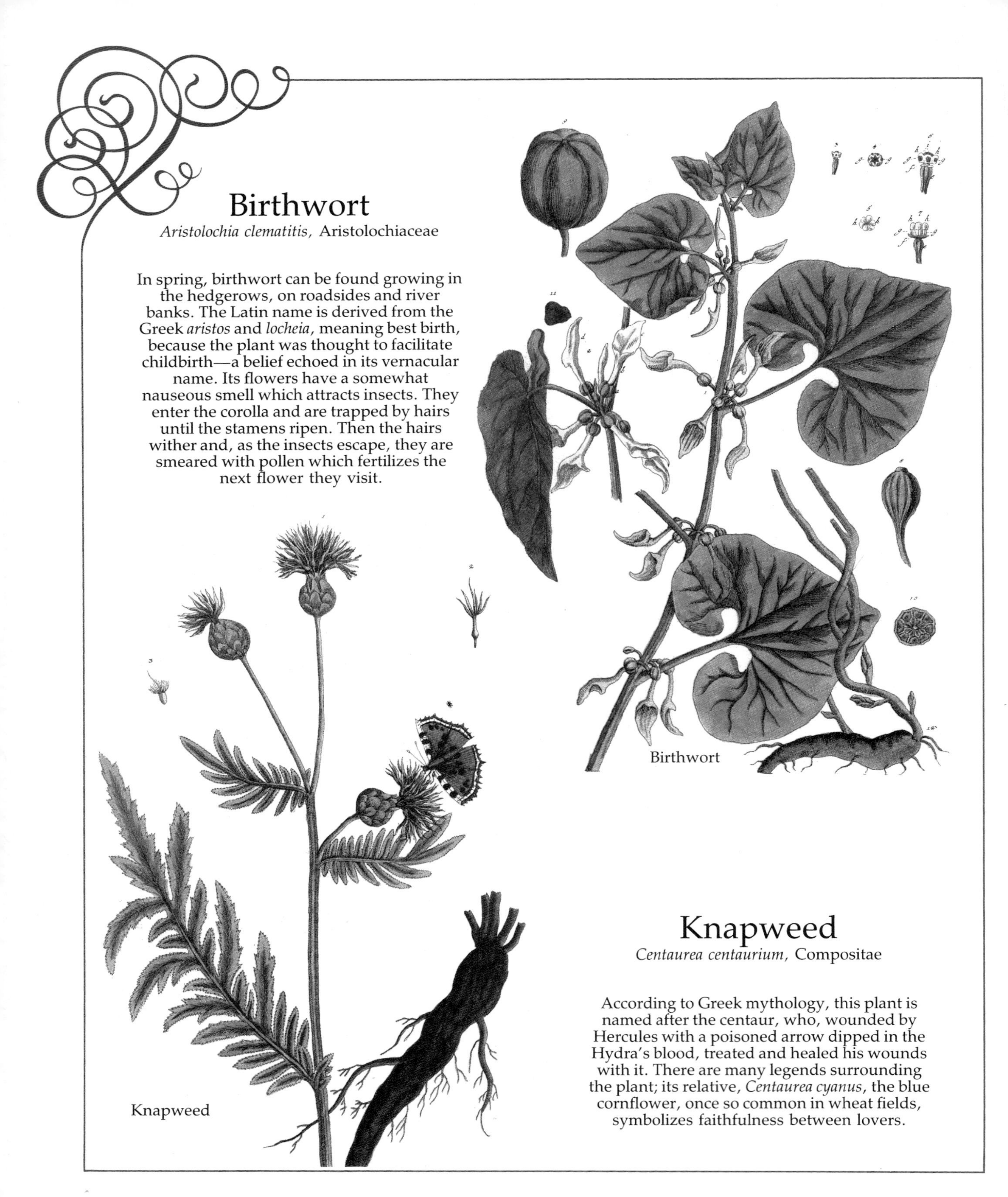

Birthwort

Aristolochia clematitis, Aristolochiaceae

In spring, birthwort can be found growing in the hedgerows, on roadsides and river banks. The Latin name is derived from the Greek *aristos* and *locheia,* meaning best birth, because the plant was thought to facilitate childbirth—a belief echoed in its vernacular name. Its flowers have a somewhat nauseous smell which attracts insects. They enter the corolla and are trapped by hairs until the stamens ripen. Then the hairs wither and, as the insects escape, they are smeared with pollen which fertilizes the next flower they visit.

Knapweed

Centaurea centaurium, Compositae

According to Greek mythology, this plant is named after the centaur, who, wounded by Hercules with a poisoned arrow dipped in the Hydra's blood, treated and healed his wounds with it. There are many legends surrounding the plant; its relative, *Centaurea cyanus,* the blue cornflower, once so common in wheat fields, symbolizes faithfulness between lovers.

Elecampane

Inula helenium, Compositae

An herbaceous plant with a thick root and showy yellow flowers, it grows wild along roads and in fields. Its purgative properties are well known, and its generic name, derived from the Greek *inaein,* means to purge.

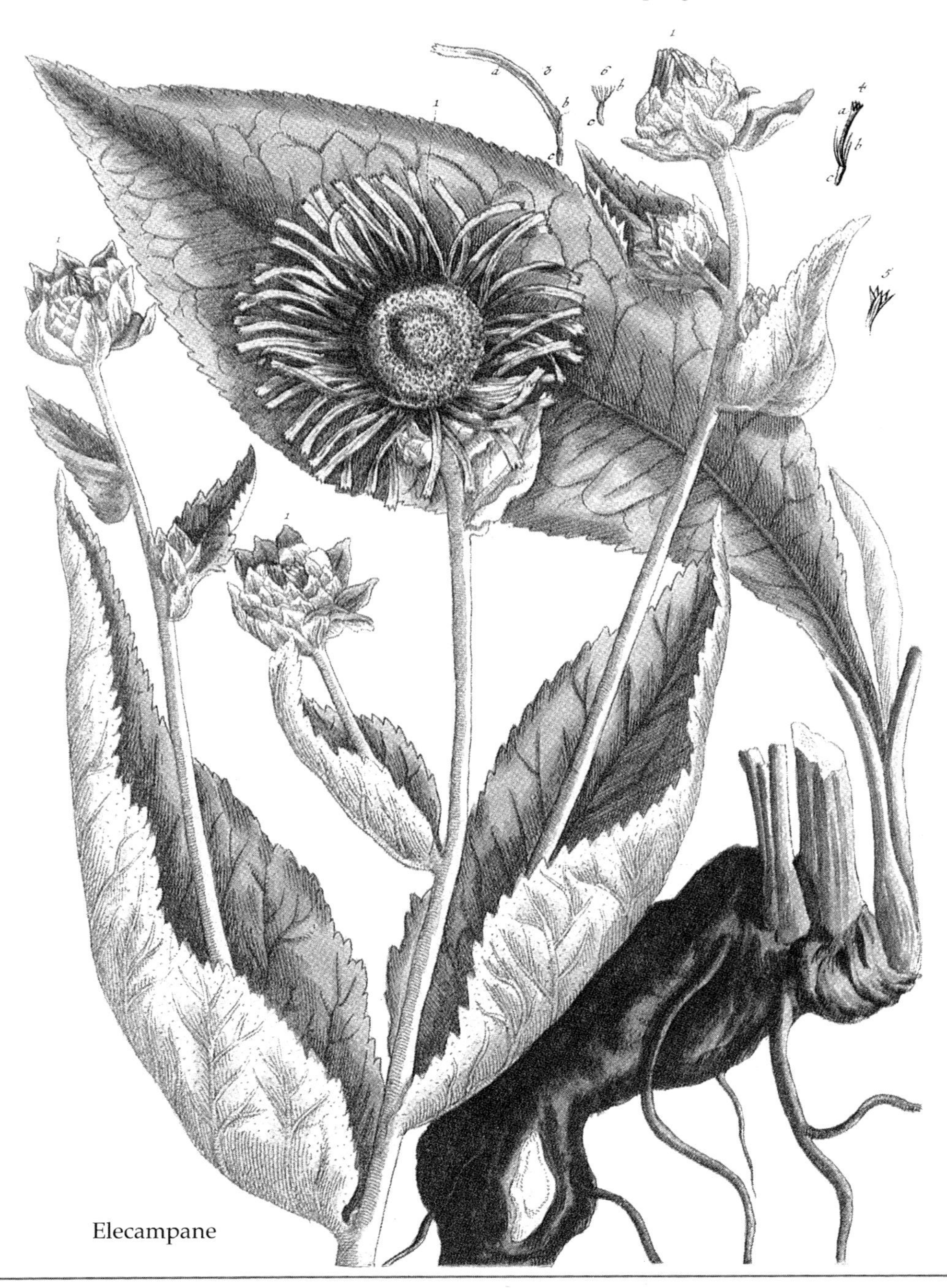

Elecampane

Hart's-tongue

Phyllitis scolopendrium or *Scolopendrium officinale*, Aspleniaceae

This plant produces neither flowers nor seeds. A fern, it reproduces itself by means of spores, hence the generic name derived from the Greek *phyllon*, meaning leaf. Its vernacular name is based on the resemblance between the leaves and a deer's tongue. Renowned for centuries for its medicinal properties, some of which have been substantiated, this plant was known to the Greek physician Galen, to the Arabs and to the Greek scholar Dioscorides. The 17th-century English herbalist Nicholas Culpeper wrote that the distilled water is very good against the passions of the heart.

Hart's-tongue

Lovage

Lovage

Levisticum officinale, Umbelliferae

The generic name probably derives from the plant which Dioscorides called *Ligysticon*, although the latter does not appear to correspond to the modern *Levisticum*. It is found growing wild in southern Europe and is cultivated elsewhere; it is mainly used as a culinary flavouring.

Aloe

Aloe vulgaris, Liliaceae

A few species of these succulents were imported by the Phoenicians from southern Africa. The aloe has had medicinal applications since ancient times. In the warmer areas of the Mediterranean the aloe has become naturalized where it has escaped from cultivation.

Burning bush

Dictamnus albus, Rutaceae

The English vernacular name of this plant derives from the fact that the essential oils exuded by the flowers can easily be set alight and will burn immediately above the bush. Its Latin name is derived from Dikté, a mountain in Crete where this shrub (*thamnos*) was common.

Red pepper

Capsicum annuum, Solanaceae

The generic name *capsicum* is derived from the Greek *kapto,* meaning to bite, referring to the hot taste of some capsicums, notably the chili pepper. *C. annuum,* the sweet red pepper, is dried and ground as the main ingredient of paprika. It has a high vitamin content and is especially rich in vitamins C and P.

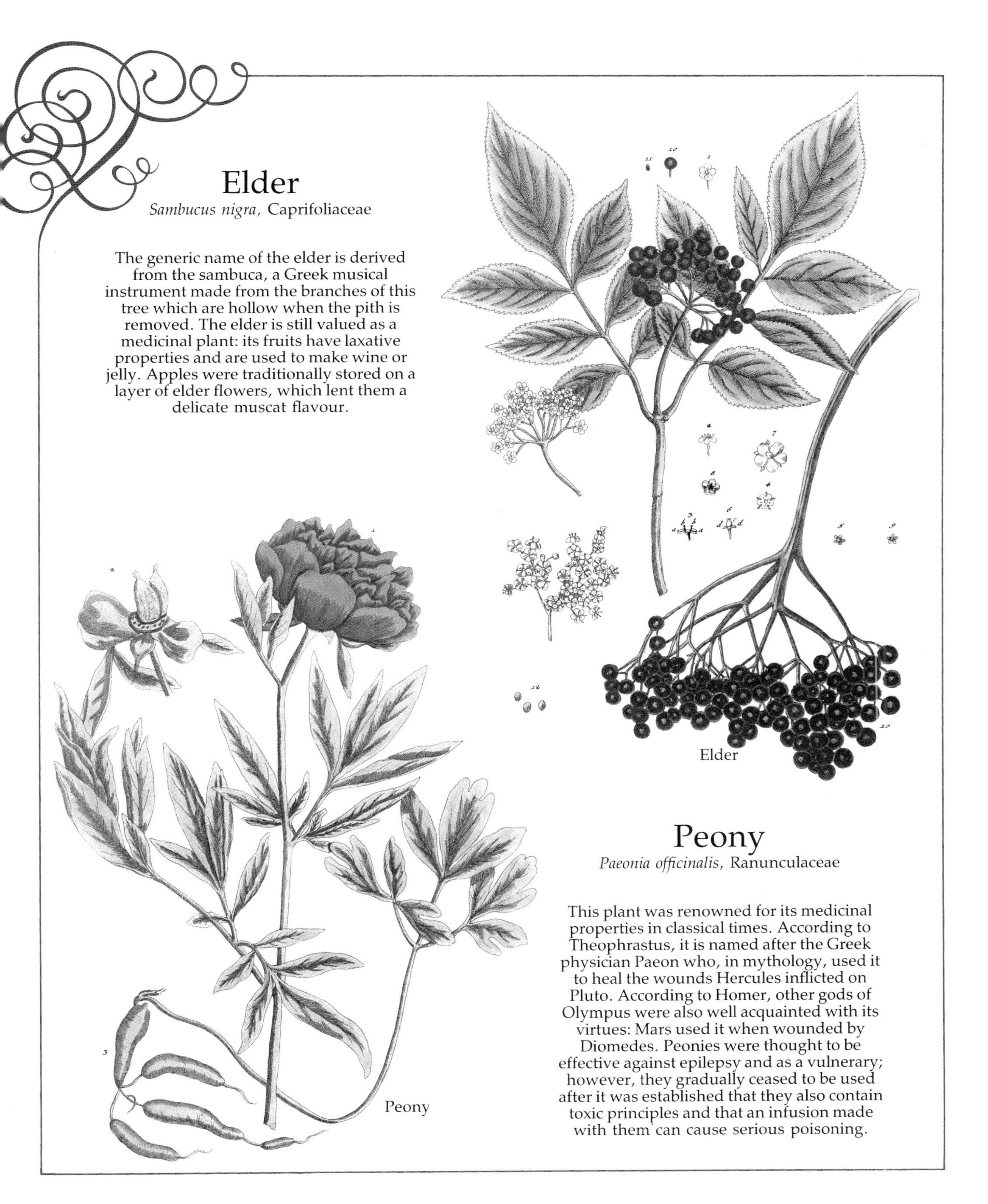

Elder

Sambucus nigra, Caprifoliaceae

The generic name of the elder is derived from the sambuca, a Greek musical instrument made from the branches of this tree which are hollow when the pith is removed. The elder is still valued as a medicinal plant: its fruits have laxative properties and are used to make wine or jelly. Apples were traditionally stored on a layer of elder flowers, which lent them a delicate muscat flavour.

Peony

Paeonia officinalis, Ranunculaceae

This plant was renowned for its medicinal properties in classical times. According to Theophrastus, it is named after the Greek physician Paeon who, in mythology, used it to heal the wounds Hercules inflicted on Pluto. According to Homer, other gods of Olympus were also well acquainted with its virtues: Mars used it when wounded by Diomedes. Peonies were thought to be effective against epilepsy and as a vulnerary; however, they gradually ceased to be used after it was established that they also contain toxic principles and that an infusion made with them can cause serious poisoning.

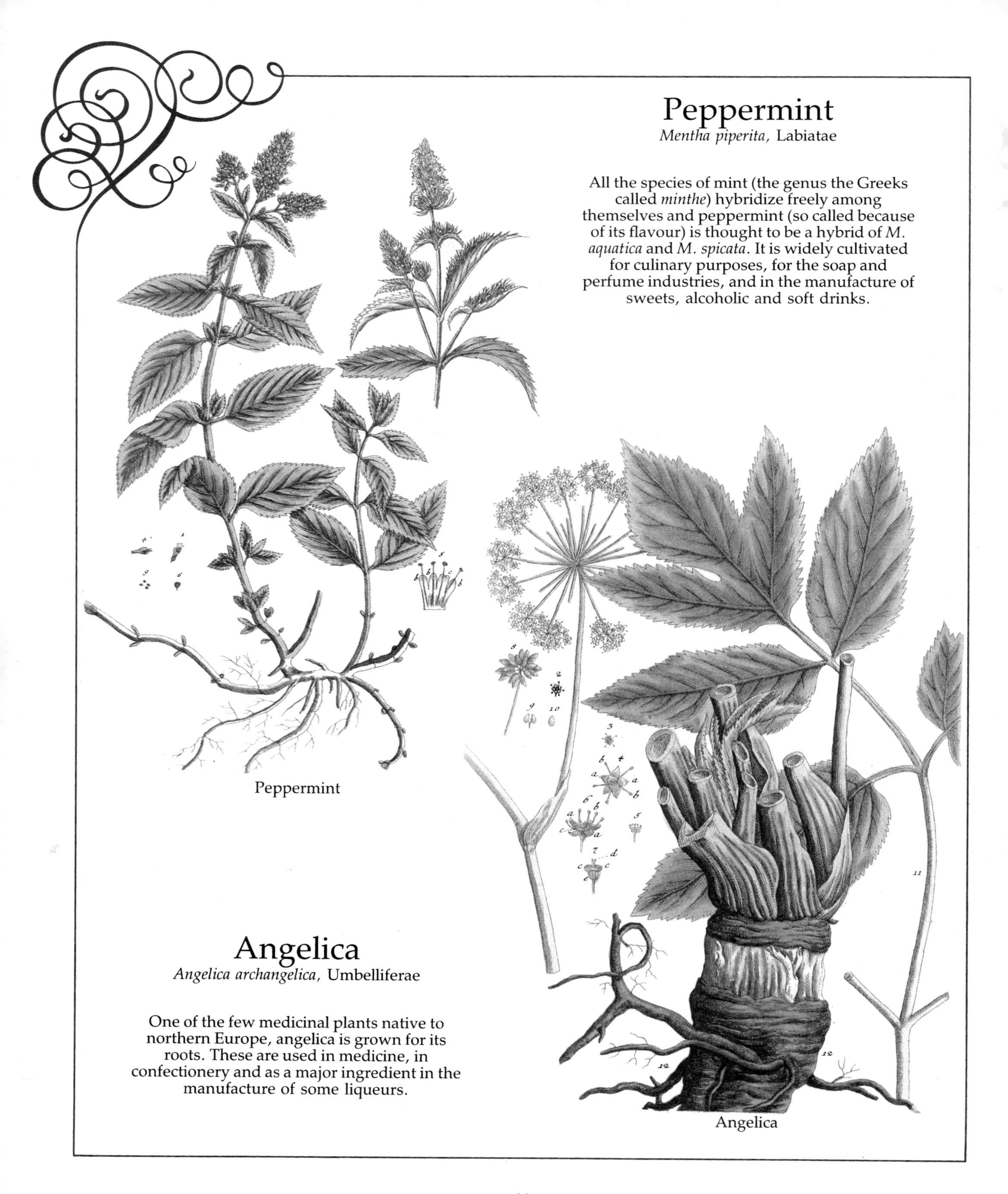

Peppermint

Mentha piperita, Labiatae

All the species of mint (the genus the Greeks called *minthe*) hybridize freely among themselves and peppermint (so called because of its flavour) is thought to be a hybrid of *M. aquatica* and *M. spicata*. It is widely cultivated for culinary purposes, for the soap and perfume industries, and in the manufacture of sweets, alcoholic and soft drinks.

Peppermint

Angelica

Angelica archangelica, Umbelliferae

One of the few medicinal plants native to northern Europe, angelica is grown for its roots. These are used in medicine, in confectionery and as a major ingredient in the manufacture of some liqueurs.

Angelica

Coltsfoot

Tussilago farfara, Compositae

The Latin name refers to the fact that even in classical times coltsfoot was used to cure coughs (*tussis*). It is one of the first plants to flower in spring. The leaves, which have a whitish, floury underside, appear later.

Coltsfoot

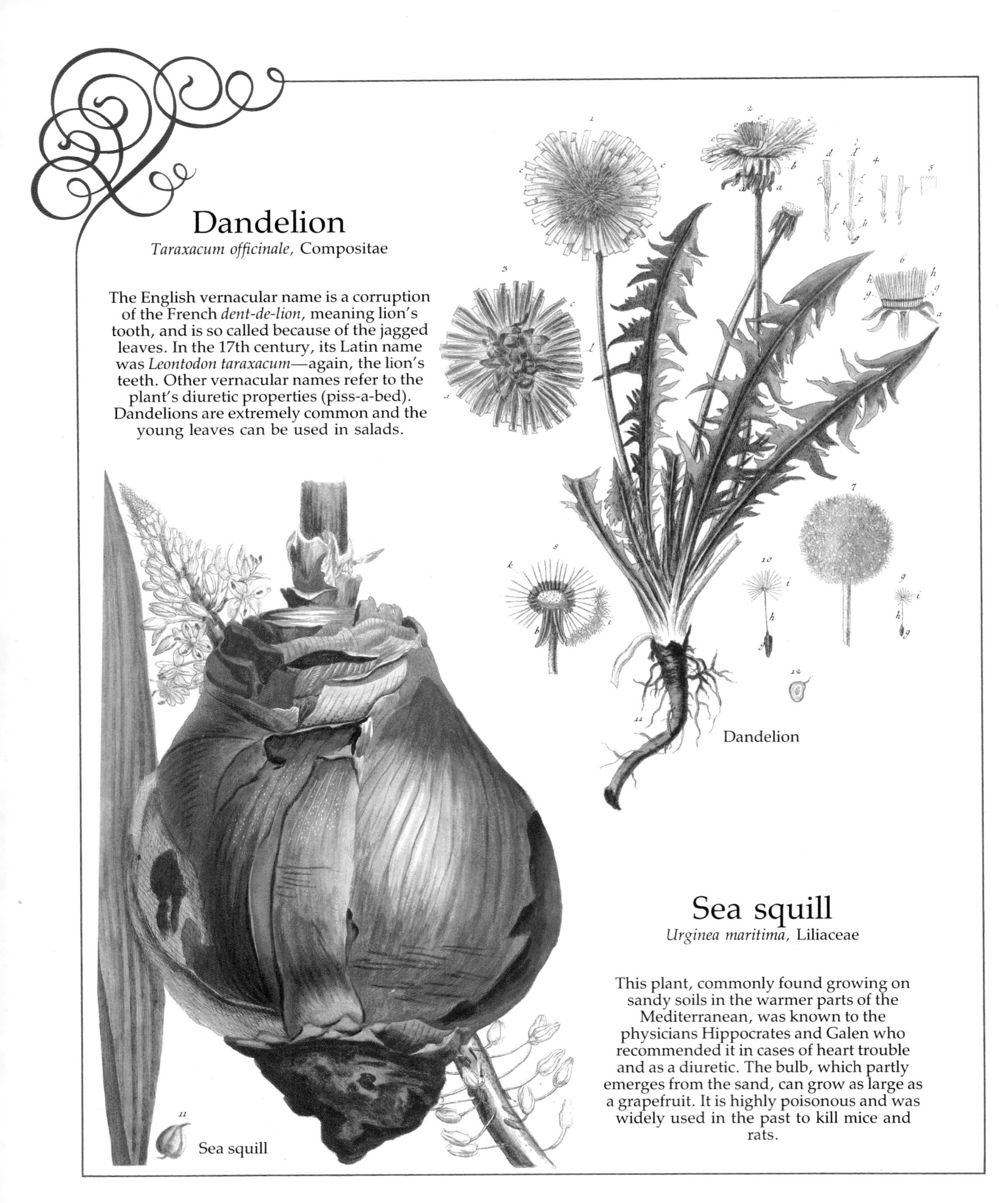

Dandelion

Taraxacum officinale, Compositae

The English vernacular name is a corruption of the French *dent-de-lion,* meaning lion's tooth, and is so called because of the jagged leaves. In the 17th century, its Latin name was *Leontodon taraxacum*—again, the lion's teeth. Other vernacular names refer to the plant's diuretic properties (piss-a-bed). Dandelions are extremely common and the young leaves can be used in salads.

Dandelion

Sea squill

Urginea maritima, Liliaceae

This plant, commonly found growing on sandy soils in the warmer parts of the Mediterranean, was known to the physicians Hippocrates and Galen who recommended it in cases of heart trouble and as a diuretic. The bulb, which partly emerges from the sand, can grow as large as a grapefruit. It is highly poisonous and was widely used in the past to kill mice and rats.

Sea squill

Belladonna

Atropa belladonna, Solanaceae

The generic name is a reference to the plant's toxicity: Atropos was one of the three Fates (the Moerae to the Greeks and the Parcae to the Romans)—the sister who administered Death by cutting the thread of Life. Belladonna, or deadly nightshade, is widespread throughout Europe and grows in shady, damp woods, particularly beech woods. Belladonna means 'beautiful woman' in Italian and the plant is so named because in the 16th century Italian women used ointments and extracts of this plant as cosmetics to whiten their skin and dilate their pupils.

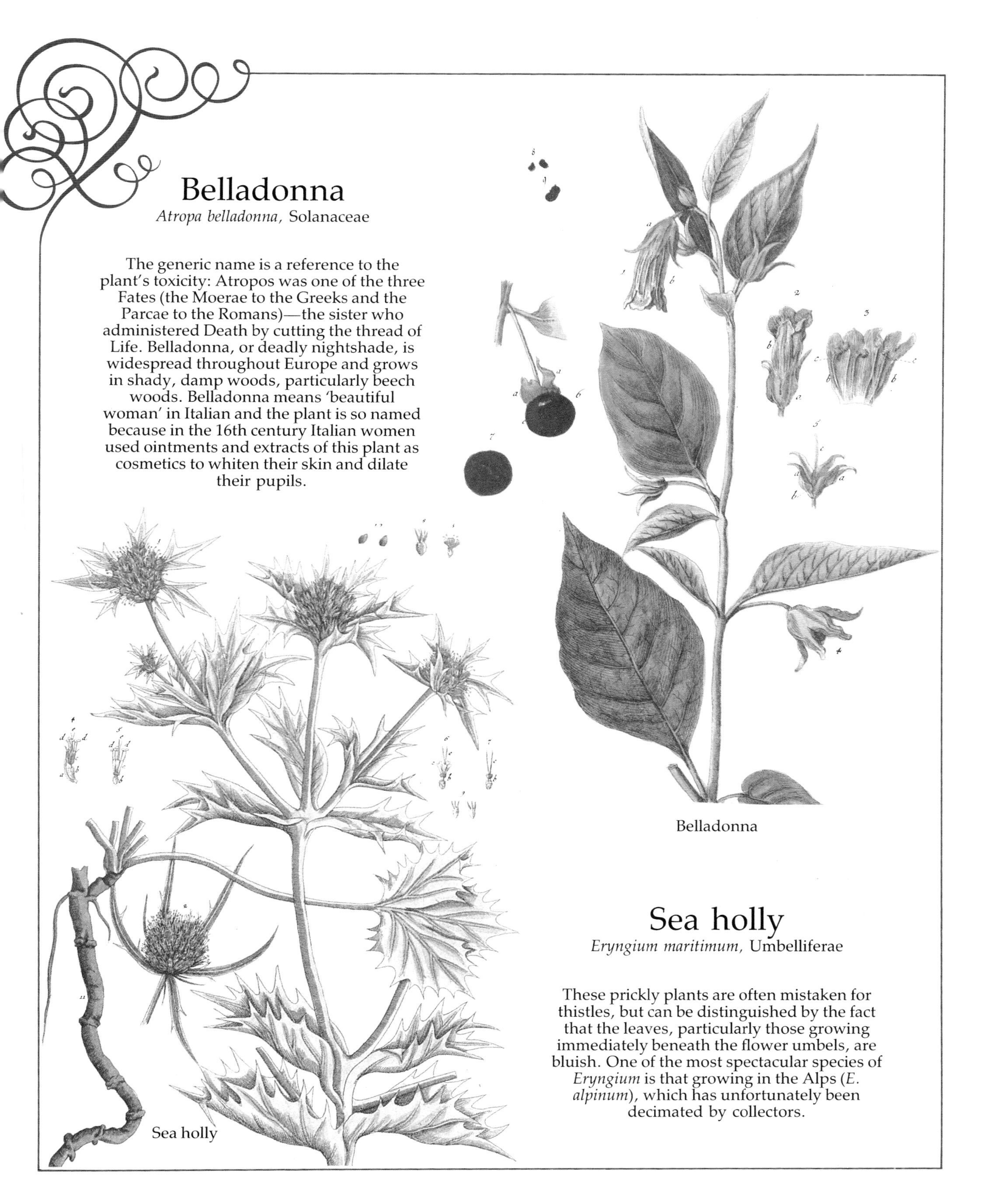

Belladonna

Sea holly

Sea holly

Eryngium maritimum, Umbelliferae

These prickly plants are often mistaken for thistles, but can be distinguished by the fact that the leaves, particularly those growing immediately beneath the flower umbels, are bluish. One of the most spectacular species of *Eryngium* is that growing in the Alps (*E. alpinum*), which has unfortunately been decimated by collectors.

Ivy

Hedera helix, Araliaceae

The generic name *Hedera* was given to this plant by the Romans. Ivy climbs a support by means of numerous short, adhesive roots borne on the stems. It is so tenacious that it sometimes kills the tree it clings to. Its typical lobate leaves grow only on the sterile branches; the fertile ones bear entire leaves. The black, shiny fruits are poisonous and, if swallowed, cause vomiting, diarrhoea and convulsions.

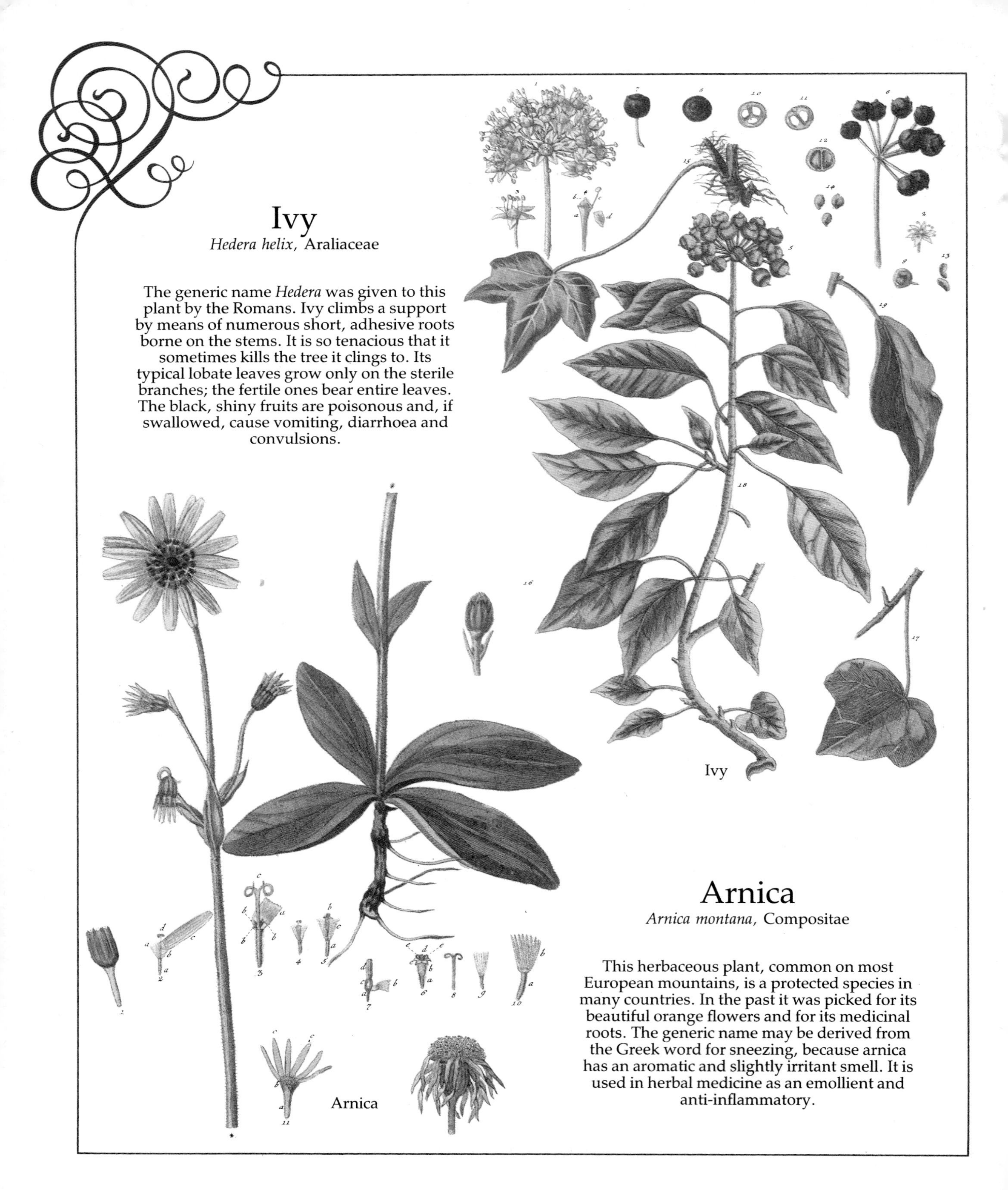

Arnica

Arnica montana, Compositae

This herbaceous plant, common on most European mountains, is a protected species in many countries. In the past it was picked for its beautiful orange flowers and for its medicinal roots. The generic name may be derived from the Greek word for sneezing, because arnica has an aromatic and slightly irritant smell. It is used in herbal medicine as an emollient and anti-inflammatory.

Butterbur

Petasites officinalis, Compositae

The generic name derives from the Greek *petasos*, meaning umbrella, because the plant has large leaves which travellers could use as umbrellas. Its common English name is butterbur because at one time the leaves were used to wrap butter.

Butterbur

Valerian

Valeriana officinalis, Valerianaceae

The Latin *valere* means to be in good health. The plant was widely used in the past and indeed infusions made from it are still taken as a tranquillizer and for nervous conditions. Plants growing in the mountains seem to be richer in active principles than those at lower altitudes, as may be inferred from the fact that they have a much stronger scent. Official medicine acknowledges the plant's cardiotonic properties.

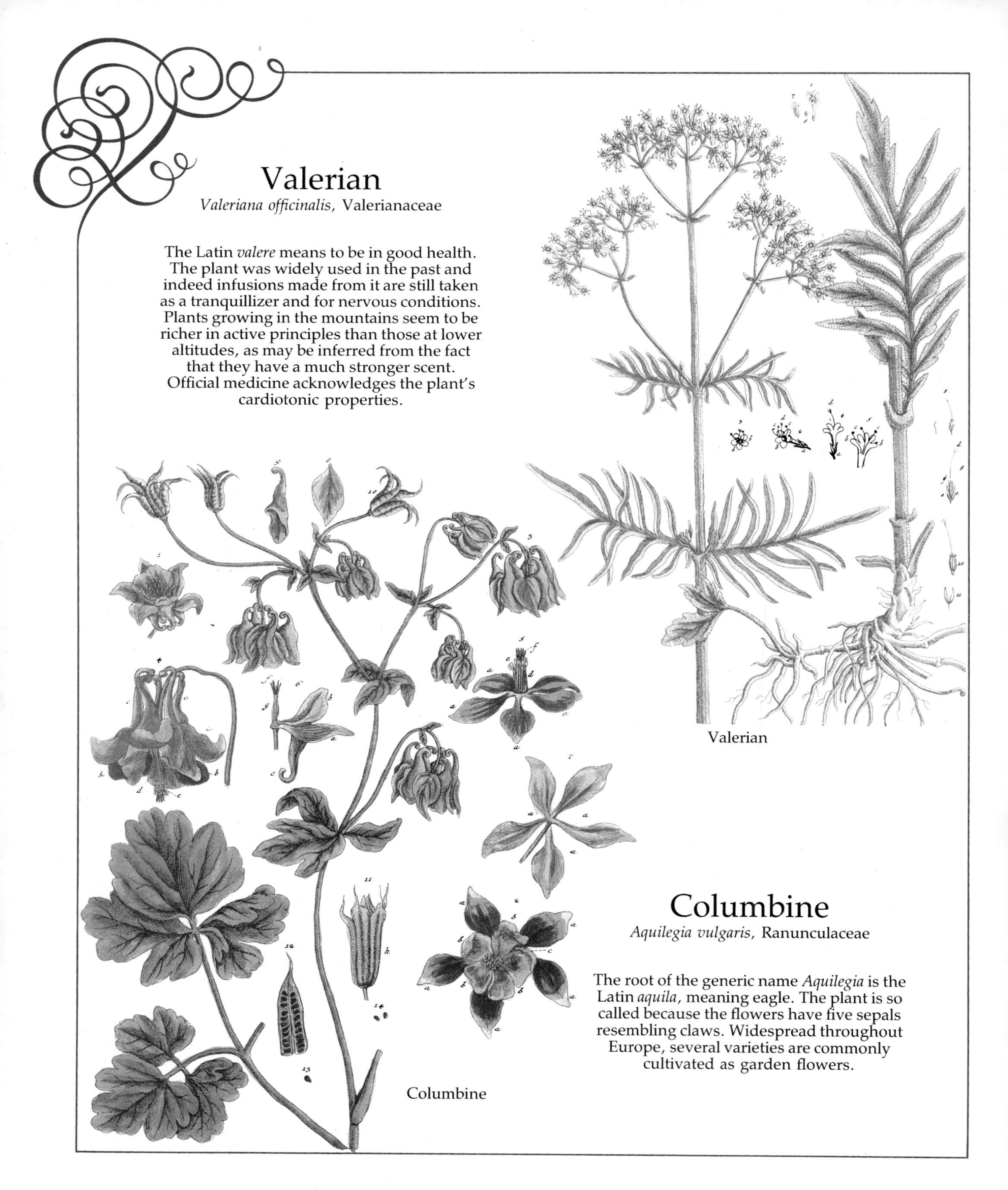

Valerian

Columbine

Columbine

Aquilegia vulgaris, Ranunculaceae

The root of the generic name *Aquilegia* is the Latin *aquila*, meaning eagle. The plant is so called because the flowers have five sepals resembling claws. Widespread throughout Europe, several varieties are commonly cultivated as garden flowers.

A knowledge of the simples

The father of the British Museum

Sir Hans Sloane was the influential physician and botanist who encouraged Elizabeth Blackwell in her unusual career as a botanical artist and whose patronage ensured her success. He was born in Killyleagh, Ireland, in 1660, studied medicine and travelled to France. In Paris he was a pupil of Joseph Pitton de Tournefort at the Jardin du Roi, and graduated at Orange. He brought a considerable plant collection back to London which was later used by John Ray, often considered the father of English botany.

Sloane was a member of the Royal Society and in 1687 he accompanied the Duke of Albemarle to Jamaica as his personal physician; the latter was to succeed Sir Henry Morgan, formerly a buccaneer, as Governor of the island. On a halt in Madeira and during the fifteen months he spent in Jamaica, Sloane collected botanical specimens avidly and returned to London with 800 new species of tropical plants, publishing a catalogue of them in 1696.

His career as a physician was also successful; Samuel Pepys was one of his patients and he encouraged inoculation against smallpox. He lived in London at 3 Bloomsbury Place, and filled the house with a large collection of plants, fossils and minerals, zoological, anatomical and pathological specimens, antiques, curiosities, prints, drawings, coins, books and manuscripts. He had also bought manuscripts by Kaempfer, a physician working for the Dutch East India Company who was the author of the first flora of Japan.

In 1717, Hans Sloane was made a baronet, the first physician to receive a hereditary title. From then on, he was heaped with honours: he was appointed President of the Royal College of Physicians, general physician to the Army, first physician to George II and President of the Royal Society in succession to Sir Isaac Newton. He was eighty-one when he retired, and two years later he moved his collections to his manor house in Chelsea.

In 1748, Sir Hans was visited in Chelsea by Frederick and Augusta, the Prince and Princess of Wales, who took such an interest in Kew. As reported in the *London Magazine*, the Prince 'took a chair and sat down by the good old gentleman some time, when he expressed the great esteem and value he had for him personally and how much the learned world was obliged to him for his having collected such a vast library of curious books and such immense treasures of the valuable and instructive production of Nature and Art'. The royal guests saw drawers 'filled with all sorts of precious stones . . . jewels polished and set after the modern fashion . . . tables spread with gold and silver ores . . . the lasting monuments of historical facts . . . brilliant butterflies . . . the remains of the antediluvian world . . . large animals preserved in the skin . . . curious and venerable antiquities of Egypt, Greece, Hetruria, Rome, Britain and even America'.

This incident is related in a brief history of the British Museum written by Marjorie Caygill, since, on his death in 1753, Sir Hans Sloane left the 79,575 items of his collection to George II in trust for the nation (in return for £20,000 to be paid to his two daughters) and this formed the nucleus of the British Museum.

Botany and Medicine

Sir Hans Sloane was first and foremost a physician, the Chelsea botanical garden was the garden of the Society of Apothecaries, and Alexander Blackwell, Elizabeth's ill-fated adventurer of a husband, was also a physician (whether a good one, it is difficult to say, but he had graduated from the prestigious University of Leiden under the famous Herman Boerhaave). It is therefore hardly surprising that the *Curious Herbal* compiled by Mrs Blackwell constitutes a delightful and discerning collection of medicinal plants. During the first half of the 18th century, medicines were still extracted from herbs just as they had been two centuries previously, when universities set up the first chairs of botany as part of their faculties of medicine.

The two disciplines were intertwined. Our modern garden plants—roses, peonies, lilies and many others—began their association with man as medicinal herbs. It is not generally realized that traces of the relationship between botany and medicine are contained in the term 'botanic' which is derived, via the medieval Latin *botanicus*, from the Greek *botane*, meaning herb. The medicinal concept is implicit. It is curious that, unlike zoology, biology and other sciences, botany does not have the ending derived from the Greek word *logia*. A firmly rooted tradition prevented 'botany' from being replaced by 'phytology', a more exact term meaning 'the science relating to plants'. A.G. Morton notes that Theophrastus (*c*.372–*c*.287 BC), the first author to write on botanical subjects whose work has survived, only used the word *botane* twice: in one instance to mean weed and in the other to refer to a plant of Gedrosia. Three centuries later another Greek author, Dioscorides, used the term with the general meaning of herb, hence our modern 'botany', although the Romans did not adopt the term and Pliny, a contemporary of Dioscorides, never mentions the word *botanicus*.

An example of plant therapy

Laurence Sterne, *The Life and Opinions of Tristram Shandy, Gentleman*, vol. VII, ch.21 (1765): 'The Abbess of Andouillettes, . . . being in danger of an *anchylosis*, or stiff joint (the *sinovia* of her knee becoming hard by long matins) and having tried every remedy:—first prayers and thanksgivings; —then invocations to all the saints in heaven, promiscuously;—then particularly to every saint who had ever had a stiff leg before her;—then touching it with all the reliques of the convent, principally with the thigh-bone of the man of Lystra, who had been impotent from his youth;—then wrapping it up in her veil when she went to bed;—then cross-wise her rosary;—then bringing into her aid the secular arm, and anointing it with oils and hot fat of animals;—then treating it with emollient and resolving fomentations;—then with poultices of marshmallows, mallows, bonus Henricus, white lilies, and fenugreek;—then taking the woods, I mean the smoke of 'em, holding her scapulary across her lap;—then decoctions of wild chicory, watercresses, chervil, sweet cecily and cochlearia;—and nothing all this while answering, was prevailed on at last to try the hot baths of Bourbon . . .'

Classical botanists and Renaissance men

Constantinople fell to the Turks in 1453; in 1529, having invaded Hungary, they laid siege to Vienna. After almost twenty years of truces and assaults, the Sultan and the Emperor Ferdinand of Habsburg signed a peace, which, although precarious, re-established diplomatic relations. The imperial ambassador to Constantinople was the Fleming, Augier Ghislain de Busbecq, a cultured man with wide interests. As a result of his diplomatic mission, the tulip, the lilac, the horse-chestnut and one of the most beautiful of all surviving manuscripts on medicinal botany were brought to Europe.

On the road from Adrianopolis the ambassador saw 'great expanses of flowers—narcissi, hyacinths and what the Turks call *tulipan*, to our great astonishment as it was the middle of winter, an unfavourable season for flowers'. Busbecq paid a considerable sum of money for tulip bulbs and had them sent to the imperial gardens in Vienna. As we have seen, it was the botanist Clusius who then imported them into the Low Countries. The name tulip, however, arises from a linguistic misunderstanding, as the Turkish word for tulip is *lalè*, while *tulipen* actually means turban. Presumably the ambassador did not follow the interpreter when the latter remarked that the flowers resembled turbans.

Constantinople was the heir to classical culture, much of which had been preserved. Busbecq copied ancient inscriptions, collected coins and bought manuscripts. In 1562 he wrote that he had seen the son of Hamon, physician to Suleiman, handling a marvellous illustrated manuscript by Dioscorides, but that he could not buy it because he had been asked one hundred ducats for it, too large a sum for his pocket. He obviously mentioned the incident in order to loosen the strings of the imperial purse. Seven years later the manuscript—known as the *Codex vindobonensis*—arrived in Vienna, where it is today, one of the treasures of the Österreichische Nationalbibliothek. It is a large book, 491 pages long, with about 400 full-page illustrations. It was written in Constantinople in AD 512, during the reign of Justinian the Great, for Juliana Anicia, the daughter of Flavius Anicius Olybrius, who reigned for a few months in 472 as one of the last emperors of the western Roman Empire, which collapsed in 476.

We know little more about Pedanios Dioscorides than what he himself tells us in his work. He was born in Asia Minor and was an army physician during the reigns of Nero and Vespasian. His *De Materia Medica* (the Latin title by which his Greek work is commonly known) is a treatise which summarizes contemporary knowledge of pharmacology. It was of fundamental importance in the development of medicine and botany, in both European and Arab culture.

Dioscorides lists and discusses about 1,000 remedies, three-fifths of which involve the use of plants. His work is based on personal experience and on material by other authors which he checked scrupulously by questioning the inhabitants of the many countries he visited in the course of his military career.

One of his sources was the *Rhizotomicon* compiled by the Greek Krateuas (2nd–1st century BC), the greatest of classical herbalists, physician to Mithridates VI Eupator, King of Pontus, sworn enemy of the Romans and himself keenly interested in poisons and antidotes. The *Rhizotomicon*, which has been lost, was a treatise on pharmacological botany. Krateuas also wrote another, more popular, work which consisted of a series of illustrations of plants, accompanied by their names and their medicinal properties. Probably in the 2nd century AD, Krateuas' illustrations were combined with Dioscorides' text, thus giving rise to the tradition of the illustrated herbal which continued until the 19th century. The influence of the original illustrations and information taken from the two Greek authors was such that they were only superseded after many centuries.

One of the illustrations in the *Codex vindobonensis* shows a lady, Epinoia (Intelligence), holding a mandrake plant so that Krateuas, seated opposite her in an elegant chair, can paint it. The 400 or so botanical illustrations of the Vienna codex are copies of Krateuas' own.

A medieval herbal published in Basle in 1542, *De Historia Stirpium*, by the Bavarian Leonhard Fuchs, again illustrates the simples being copied from nature: a specimen of *Agrostemma githago* is being drawn by Albert Meyer while Heinrich Füllmaurer engraves a drawing on a wood block.

It has probably been of crucial importance for the history of botany that the six books of Dioscorides' *De Materia Medica*—a practical rather than a scientific work—have been passed down through the centuries while the writings of that other great classical writer on botany, Theophrastus, were lost and only rediscovered in the Renaissance. Theophrastus was born Tirtanus, at Eresos, on the island of Lesbos; he was renamed by Aristotle, his teacher and colleague at the Athenian Lyceum—Theophrastus means 'divine orator'. He wrote two works which are known by their Latin titles, *Historia plantarum* and *Causae plantarum*, and which faithfully reflect his own teaching at the Lyceum *c.*300 BC. At this time Greece was divided into kingdoms formed after the conquests of Alexander the Great. The Lyceum was the fee-paying school where the ruling classes were educated.

Theophrastus' approach to botany was pragmatic and he analysed the problems posed by the plant world with a scientific clarity which was later lost. His two treatises were translated into Latin in 1483 by Theodore Gaza, a Greek born in Salonica, who fled to Italy from the Turks. Gaza taught Greek in Ferrara and philosophy in Rome. The publication of his translation is important in the history of botany as a science, since Theophrastus' work thus became available to scholars. However, Dioscorides continued to be the unchallenged authority for a long time.

Busbecq's correspondence contains references to the fact that he sent botanical information, drawings of plants preserved in his Constantinople study and two manuscripts by Dioscorides (not, however, the *Codex vindobonensis*) to Pierandrea Mattioli, the Siennese physician at the imperial court in Vienna. Mattioli is the best known of Dioscorides' commentators; he discussed the classical text at length, with annotations which far exceeded the original propositions. He added plants which he had either collected personally in the various places where he lived or which he had been sent by colleagues. One of these plants was the sunflower. Mattioli also drew liberally on the work of his friend Luca Ghini, who founded the botanical garden at Pisa.

The son of a doctor, Mattioli practised medicine in Siena and in Rome until the town fell to Charles V in 1527, when he moved to northern Italy. In 1544 his *Libri cinque della historia et materia medicinale* (Five books on the history and subject of medicine) was published in Venice. This was the first edition of his *Commentarii* on Dioscorides, which was later revised, reprinted and enlarged many times and translated into several languages. It was a great success, with some fifty editions and 32,000 copies sold of the first few impressions.

On the strength of his scholarly reputation, Mattioli was summoned to Prague and became court physician to Ferdinand I and Maximilian II. When he retired he moved to Trento and died there soon afterwards of the plague, in 1577, at the age of seventy-six.

The *Codex vindobonensis* had been annotated by a Turk, and Turkish and Arab plant names had been added in the margin. Hamon, Suleiman's physician, had then added the Hebrew equivalents. To relate the ancient plants accurately to those known in the 16th century posed a serious problem, especially since Mattioli's purpose was to enable the therapeutic properties discovered by the Greeks to be used. He is accused of having carried out experiments on prisoners to find out whether *Aconitum napelles* was the most poisonous plant of its genus. We now know it is deadly poisonous. The first edition of his *Commentarii* was not illustrated; later editions included woodcuts by Giorgio Liberale of Udine, and by the German Wolfgang Meyerpeck.

A 17th-century physician

Mattioli's aim, therefore, was to arrive at 'the true knowledge of the simples'. Until a century ago, pharmacology was still oriented almost entirely towards the use and production of medicinal herbs. Modern synthetic drugs have not altogether replaced these herbs. They still represent about a quarter of the basic ingredients of the pharmaceutical industry. Medicines used to be termed 'galenicals' after Claudius Galen (AD *c*.129–*c*.201), the Greek physician who treated the Emperor Marcus Aurelius; Galen's bust, together with that of his fellow Greek Hippocrates (*c*.460–*c*.377 BC), always graced the shelves of old apothecaries' shops. A galenic preparation contained one or more organic ingredients obtained by some natural or mechanical process such as grinding, extracting, percolating, infusing, or mixing natural products.

But what did the practice of medicine really consist of in the days of Trew, Ehret and Mrs Blackwell? We are given some indication by what we know of those who have, for one reason or another, become part of the history of medicine. All that remains of their less worthy colleagues are the unflattering caricatures of Molière and Lesage. One of the best 17th-century physicians was the Englishman Thomas Sydenham. His reputation as a doctor was well established when he was paid a visit by Sir Hans Sloane. The latter's letter of introduction commended his own academic achievements and his knowledge of anatomy and botany, but Sydenham received him coolly and read him a lecture on the following lines. A knowledge of anatomy and botany counted little or nothing as far as practical medicine was concerned—Sydenham knew an old woman in Covent Garden who probably knew more about botany than Sloane, and as for anatomy, his own butcher was hard to beat, able as he was in cutting up quarters of beef. He urged the young man to sit at a patient's bedside, as only there would he learn something about sickness.

Sydenham was a very successful doctor, probably because he insisted upon clinical observation of his patients and their symptoms. He believed that acute illnesses were due to an inflammation of the blood which provoked a spontaneous reaction in the organism. He therefore had great faith—like Hippocrates—in the healing power of nature. Two drugs made an important contribution to his fame: quinine and 'Sydenham's liquid'. The latter is in fact laudanum, a tincture of opium with wine or alcohol and other ingredients. It was widely used, prepared in liquid form according to Sydenham's recipe, and is still in use today. One of the formulas adopted by modern pharmacology is as follows: 15 parts opium, 70 parts alcohol at 60°, saffron, cinnamon, cloves and water. This type of laudanum should not be confused with what was called mineral laudanum: mercuric chloride (corrosive sublimate). Medicinal laudanum is an anti-spasmodic and a pain-killer.

Extracts of mandrake, valerian, belladonna and Indian hemp have been used to relieve pain since antiquity. According to Dioscorides, 'mandrake roots should be boiled until the liquid is reduced by a third. The decoction should be administered to the person to be operated upon or cauterized, so as to make him insensitive to pain'. The Greek word used by Dioscorides for insensitivity is *anaesthesia*, a term reintroduced by the American doctor Oliver Wendell Holmes when the use of ether in operating theatres became widespread in the mid-19th century. Mandrake contains alkaloids similar to those of the opium poppy.

Laudanum was used a great deal by another English doctor, John Brown (1735–88), who believed that life is determined by stimuli which may be internal (those which influence the senses, muscular contractions, the activity of the brain which manifests itself as thoughts and emotions) or external (the air we breathe, the food we eat, the temperature of our environment). Health is the right balance between these stimuli; illness consists of either too much excitation (a 'stenic' illness) or of too little (an 'astenic' illness). Treatment should therefore be based either on sedatives or stimulants. Brown himself used to take fifty drops of laudanum in a glass of whisky and died an alcoholic at the age of fifty-three.

The properties of quinine

Quinine, like ipecacuanha, sarsaparilla and other medicinal drugs (the word drug is probably derived from the Dutch *drog*, meaning dry, dried matter) is a gift of the Americas, or, if one prefers, a consequence of Christopher Columbus's thirst for gold and knowledge. We have already seen that John Evelyn mentioned 'the Jesuit's bark' after one of his visits to the Chelsea botanical garden. Apparently, it was discovered when a Jesuit missionary to Peru was cured of fever by a concoction administered by a local medicine man. The latter refused to reveal the recipe for the potion, but the Jesuit managed to discover that the active ingredient was the bark of a plant native to the Andes.

There are other versions of the story, but the Jesuits certainly played an important role in spreading the use of the drug. In Rome it was opposed by the professional physicians but promoted by the Church, so much so that when it reached England it was scorned as a 'popish' remedy and was not administered to Cromwell—who quietly died of malaria. During the Restoration, however, the Jesuit's bark enabled Robert Talbot to win fame and fortune. This apprentice chemist used a concoction of rose petals, lemon juice, water and quinine to cure Charles II of a fever; the King then made him Royal Physician and knighted him. Talbot at first kept his recipe secret, but in 1681 he sold it to Louis XIV, the Sun King, for 3,000 louis d'or with the promise that it would not be revealed during his lifetime. He died the following year.

Quinine is an alkaloid found in the bark of certain evergreen members of the Rubiaceae, a family to which the coffee plant and the gardenia also belong. According to another version of its discovery, the Spanish magistrate of a small town in present-day Ecuador, near the Peruvian border, was the first to test the effectiveness of the febrifuge in 1630; seven years later he sent it to the Countess of Chinchon, wife of the Viceroy of Peru—hence the drug's other name of 'Countess's powder'. This story has not been borne out by scrupulous examination of contemporary archives but it is true that the Count of Chinchon introduced the bark into Spain. Linnaeus called the species *Cinchona officinalis* in his honour, albeit misspelling the Viceroy's name. The name quinine is also a derivation of Chinchon.

In the 18th century the French botanist Joseph de Jussieu took part in a scientific expedition to Peru organized by Charles Marie de La Condamine (a young officer and friend of Voltaire's), and unsuccessfully tried to establish cinchona plants in Europe. In the first half of the following century the Dutch succeeded in cultivating a few plants on Java. In 1859 the British organized an expedition to collect plants in the Andes and, overcoming the difficulties raised by local authorities, exported plants and seeds to Ceylon and southern India. As a result, quinine, so valuable in the fight against malaria, the scourge of many countries, became widely available at a very low cost. In 1898 Giovanni Battista Grassi showed that malaria is caused by a parasite of the *Anopheles* mosquito, which acts as the carrier; only after the Second World War was the disease effectively eradicated.

Superstitions and warnings

The roots of peonies should be dug up only at night; if done by day a woodpecker would peck out the eyes of the person attempting it. In the 17th century it was believed that mandrake grew from the drops of blood which fell onto the soil from a hanged man left on the gallows. Three circles should be drawn around the plant with a sword before digging it up. According to Shakespeare, mandrakes shriek when torn out of the earth, 'that living mortals hearing them run mad' (*Romeo and Juliet*). Hellebore must be uprooted with one's eyes fixed on the east and only when no eagles are flying overhead —a certain sign of death for the herbalist.

The Romans believed that their god of medicine, Aesculapius, sent the melancholic and the mentally deranged to Antycira in the Gulf of Corinth, since the best hellebores grew there. The belief that the black hellebore was beneficial in treating mental illness dates back to the Egyptians, and was shared by the Greeks.

The grammarian and mythographer Apollodorus of Athens (2nd century BC) recounts the myth of Melampus and the daughters of the King of Argos. The daughters of Proetus and Stheneboa—Lysippe, Iphinoe and Iphianassa—were driven to madness for not having observed Dionysus' festival. They raged and wandered through the lands of Argolia, in Arcadia and in the Peloponnese, where they frequented the wilderness and behaved in the strangest and most shameless manner. Melampus, the son of Amitaon, was a renowned soothsayer and the first to discover the medicinal secrets of potions and purges; he offered to restore the girls to sanity in exchange for one-third of Proetus' kingdom. At first the King refused, but his daughters' madness grew worse and other women, affected by it, left their homes and families and hid in the wilderness. At last Proetus had to turn to Melampus, who, in the meantime, had increased his price: another third of the kingdom was to be given to his brother, Bias. Iphinoe died but the other two girls were healed and married the two brothers. According to Dioscorides' account of the myth, written in the 1st century AD, the healing medicine was black hellebore.

Aulus Cornelius Celsus, a contemporary of Dioscorides, writes in the medical volumes of his *De re medica* that when people suffer from hallucinations, the first thing to note is whether the latter are happy or unhappy. Unhappy hallucinations should be treated with black hellebore (*Helleborus niger* or Christmas rose, a member of the

Ranunculaceae family) whilst happy hallucinations should be treated with white hellebore (*Veratrum album*, Liliaceae). He adds a reminder that 'happy madness is less serious than unhappy melancholy'. Both plants are poisonous, like many other medicines, but hellebore has played a sinister role in history.

Prospero Alpino, the 16th-century Italian botanist and physician, took issue with the priests who used it everywhere 'to treat chronic infirmities, in the belief that they were caused by witchcraft and therefore curable only with sacred words, sacred ointments and other medicines invented by themselves. Thus they prescribe oils to be taken orally which contain the root of white hellebore and pass them off as holy . . . Such medicines, taken orally, can stifle even the best of spirits and kill the patient.' Not until the end of the 18th century do we read in the more humane works of Philippe Pinell, the founder of scientific psychiatry, that 'one should not mourn the demise of this remedy, since its prescription was based on blind empiricism, without any solid foundation such as knowledge of the history of the patient's symptoms and of the various types of mental aberration'.

In the first century AD, Pliny wrote in his *Historia Naturalis*: 'Experience, the most effective teacher in all things and above all in medicine, gradually degenerated into mere words and stories. It was in fact pleasanter to listen attentively in a classroom than to go out into the fields during the different seasons of the year to look for the various plants.' An 18th-century Frenchman seems to echo Pliny when he writes: 'I know from everyday experience that most herborists only know a few of the plants which country people bring to them at the right time of year. In so far as they distinguish them, they use corrupted names and, by mixing up the various species, cause great damage both to the health of their patients and to the reputation of physicians and chemists' (Pierre Jean-Baptiste Chomel, physician and botanist, 1671–1740).

An enlightened approach to medicine

The Milanese Count Alessandro Verri, author of *Commentariolo sulla ragione umana*, (A commentary on human reason; 1766), wrote in a contemporary periodical: 'I regret that these are no longer the days of Martianus Capella, when, as he assures us and we can well believe him, fevers were cured by music, and Asclepiades of Bithynia cured deafness with the sound of a trumpet. I regret that these are not the days of the Cretan Taletas and of Terpandrus: the former banished the plague from Sparta with the music of his lyre, and the latter quelled an insurrection there by the same means. I am sorry that gout can no longer be cured with a light tune on a flute as Theophrastus maintained, and that the time of Democritus is over, when the pipes were played to extract the viper's venom from the blood. Nowadays quinine is taken to drive away a fever, and no doctor employs trumpets to treat deafness, as, according to modern thinking, this would worsen the complaint. As for the plague, it only abates when it has run its course; it is kept at bay through the efforts of the lazar houses and the quarantine hospitals. Gout is avoided through exercise and diet, and cured only with patience; the viper's bite has its own remedies which sometimes work and more often do not. Our Boerhaave, our Redi, our Sauvage, our Tissot, our Haller, know nothing of musical medicine.'

Of the physicians mentioned by Count Verri in this quotation, the first had been dead for almost thirty years: the Dutchman Herman Boerhaave (1668–1738) was an influential Professor of Medicine and Botany at Leiden, a university renowned for its unprejudiced and pragmatic ideas. Histories of medicine regard him as one of the founders of modern scientific medicine because his teaching methods were based on observation of the patient and knowledge of his clinical history. He carried out autopsies in the anatomy theatre in front of his pupils, and adopted a materialistic interpretation of human physiology. His father was a clergyman who would have liked his son to follow in his footsteps. Boerhaave therefore studied theology, philosophy and oriental languages, while teaching mathematics to support himself. His true calling, however, was science.

In 1709, after being a lecturer for eight years, he was appointed Professor of Medicine and Botany. Medical theory at that time was divided into the two schools of iatrophysics and iatrochemistry, which tended to rely on physics or chemistry respectively, in order to explain physiological phenomena. Boerhaave was an eclectic—a wise position to take, since physics was more advanced than chemistry, but an understanding of the latter was to be crucial in the development of medicine. The Chair of Medicine included the teaching of botany and carried with it overall responsibility for the botanical gardens. As a student, Boerhaave collected plants in the polders of his homeland but had avoided the 'boredom' of botany lectures. He now turned his scientific and pragmatic mind to this field as well; during the first ten years of his professorship, some 2,000 new plants were added to the university's botanical gardens, partly as a result of the help of the officials of the Dutch East and West Indies Companies, and their contact with tropical Asia and America.

Boerhaave thought medicine should be linked to the natural sciences and was thus able to approach the scientific problems of botany with exceptionally wide knowledge. Linnaeus quotes a simple and clear definition he gave of botany as 'that branch of natural science which enables us to get to know and to remember plants with great pleasure and a minimum of effort', by teaching us about their structure, properties and applications. This concept of a general natural science illustrates the change in the relationship between botany and medicine that had taken place since the time of Luca Ghini and Pierandrea Mattioli.

Casanova visits an eminent scientist

'Mr Haller was a man six feet high, broad in proportion, he was a well-made man and a physical as well as mental colossus. He received me courteously and when he had read M. de Muralt's letter, he displayed the greatest politeness, which shows that a good letter of introduction is never out of place.' Casanova visited Haller, another of the physicians mentioned by Count Verri, in 1760. The portrait he painted of him in his *Mémoires* (translated by Arthur Machen, 1959) throws some light on the cultural characteristics of an 18th-century scientist.

Albrecht von Haller (1708–77) was born in Berne and was a brilliant pupil of Boerhaave, who was forty years his senior. He became a highly esteemed physiologist and was a friendly correspondent of Christoph Jakob Trew. He had been a child prodigy and had compiled a Greek-Hebrew dictionary at the age of nine. He studied at Tübingen, Leiden (where he graduated before he was twenty), London, Paris and Basle. He taught anatomy, botany and medicine at Göttingen for seventeen years and then returned to Switzerland.

'This learned man', Casanova goes on, 'displayed to me all the treasures of his knowledge, replying with exactitude to all my questions, and above all with a rare modesty which astonished me greatly, for whilst he explained the most difficult questions, he had the air of a scholar who wanted to know; but on the other hand, when he asked me a scientific question, it was with so delicate an art that I could not help giving the right answer.'

On the subject of Latin he remarks, 'Haller had written to Frederick the Great that a monarch who succeeded in the unhappy enterprise of proscribing the language of Cicero and Virgil from the republic of letters would raise a deathless monument to his own ignorance. If men of letters require a universal language to communicate with one another, Latin is certainly the best, for Greek and Arabic do not adapt themselves in the same way to the genius of modern civilization, being much further removed from the spirit of the modern age.'

Casanova then proceeded to ask Haller's opinion of some leading contemporary figures: 'During dinner I asked if M. de Voltaire came often to see him. By way of reply he recited these lines of the poet: "*Vetabo qui Cereris sacrum vulgarit arcanum sub iisdem sit trabibus.*" (Horace: I shall not allow the violator of Ceres' sacred mystery to sit under the same roof) [undoubtedly a reference to Voltaire's attacks on the Church] . . . I believe that in matters of that kind [religion] M. Haller judged only by his heart. I told him, however, that I should consider a visit to Voltaire a great event, and he said I was right. He added, without the slightest bitterness, "M. de Voltaire is a man whom it is worth knowing, although, in spite of the laws of nature, many persons have found him greater at a distance than close at hand." . . . I had been reading at Berne Rousseau's *Héloïse*, and I asked M. Haller's opinion of it. He told me he had once read part of it to oblige a friend, and from this part he could judge the whole. "It is the worst of all romances because it is the most eloquently expressed. You will see the country of Vaud, but don't expect to see the originals of the brilliant portraits which Jean-Jacques painted. He seems to have thought that lying was allowable in a romance, but he has abused the privilege. Petrarch was a learned man, and told no lies in speaking of his love for Laura, whom he loved as every man loves the woman with whom he is taken; and if Laura had not contented her illustrious lover, he would not have celebrated her." Thus Haller spoke to me of Petrarch, mentioning Rousseau with aversion. He disliked his very eloquence, as he said it owed all its merits to antithesis and paradox.'

The Doctrine of Signatures

We opened this chapter with a quotation from Shakespeare. One of the sources of *Much Ado About Nothing* may have been a comedy by the Neapolitan Giambattista della Porta (1535–1615), an extraordinary character who was a physicist, the author of works on agriculture and on magic, a traveller, playwright and probably a charlatan. In 1588 his *Phytognomonica VIII libris contenta* (Treatise on plants in eight volumes) was published in Naples. He was one of the so-called 'mystical' botanists, a follower of the Doctrine of Signatures first expounded by Paracelsus.

According to this theory nature indicates a plant's therapeutic value by means of a *signatura*, a recognizable similarity between certain plants and the organ they are supposed to cure. Thus the value of *Hepatica* species (liverwort) in treating liver afflictions is indicated by the fact that the leaves are shaped like the liver; peony is good for the brain because its pistil resembles this organ; St John's wort is used in the treatment of wounds since its leaves are riddled with glands. Paracelsus' followers developed this thesis: since strawberries resemble the growths caused by leprosy, they must be a remedy for it, and it is obvious that the walnut is suitable for all brain complaints—its shell is the skull, the inner skin the meninges and the nut the brain itself.

From his study of plants, Giambattista della Porta reached the following conclusions: any root or fruit shaped like a heart is a cardiac drug; the maidenhair fern is effective against baldness; plants with seed-pods are an antidote to the poison of scorpions; scaly fruits, like pine-cones, are good against toothache. Finally, he compared the physiognomy of plants, particularly their flowers, with the organs of animals, and from the temperament of the latter he deduced the plant's therapeutic strength.

The author of the Doctrine of Signatures, which was only of secondary importance in his work, was a gifted man with many revolutionary and controversial ideas. Theophrastus Bombastus von Hohenheim (1493–1541), known as Paracelsus, was born in Einsiedeln, Switzerland. He cured the Humanist Johann Froben (Frobenius) of a wound in the leg, which other physicians had recommended be amputated. Frobenius mentioned his cure to his friend Erasmus, who was also in need of treatment. Erasmus wrote to Paracelsus, who sent him the necessary medicines, and he too recovered. The town council of Basle then conferred upon Paracelsus the Chair of Medicine, although it is not certain whether he did in fact graduate in medicine at Ferrara, as he claimed. He immediately caused a scandal by lecturing in German rather than in Latin, as was universal practice; he also publicly burnt Galen's works, which were, and indeed continued to be, the Bible of medical science, and he openly attacked his colleagues and established medical opinion. Two years later he was expelled.

Although an effective, practical physician, Paracelsus divided his energies between the study of nature on the one hand and magic, the occult, alchemy and astrology on the other. It is possible to detect in his work the seeds of chemotherapy, the treatment of disease with chemicals, which eventually superseded herbal pharmacy. He led an itinerant life, but wrote a great deal; when asked how he had managed to collect the material for his many works, he answered: 'I searched for my art, often risking my life. I have never been ashamed to learn from vagrants, executioners, barbers, whatever seemed useful to me.' He also maintained that the physician's character could affect the recovery of a patient more than any medicine. He tried out the effects of belladonna, opium and stramonium on himself and died as a result.

Preparing the simples

The use of the term 'simple', meaning a medicinal herb, is misleading in that such plants are made up of a multiplicity of compounds. It is in these compounds that the plants' active principles lie, and a herbalist therefore has to be able to separate them by a suitable method. Apart from grinding the simples in a mortar, one of four basic procedures may be carried out—decoction, infusion, maceration and extraction of the juices. The resulting preparations are called, respectively, decoctions, infusions, dilutions and extracts.

The high temperature required for decoction can modify some of the plant's compounds and the naturally volatile parts of the active principles are lost. This method is used mainly to obtain medicines from the woody parts of a plant. It involves boiling the roots or other suitable parts in water, for a length of time determined by herboristic practice, leaving them to macerate, and then filtering the liquid.

Infusion is a suitable method for the soft

and fragile parts of a plant, such as the leaves, buds or flowers. The simples, which should have been dried, are placed in glass or earthenware—never metal—containers. Boiling water is then poured over them and the container covered immediately in order to conserve the volatile principles. After it has been left for a prescribed amount of time, the infusion is filtered.

Maceration is the simplest process. The therapeutic plant or a part of it is placed in cold water. Obviously the resulting liquid will only contain the principles which are soluble in water.

Juices are obtained from fresh plants which have a high water content; they are immersed briefly in boiling water and are then passed through a press. An extract is obtained from the concentrated juices. A liquid, usually a mixture of water and alcohol or ether, is added to the juices, which are left to evaporate. The extracts thus obtained are, respectively, hydro-alcoholic or ethereal extracts. Porcelain containers which can withstand high temperatures should be used. Depending on the degree of evaporation obtained, the extracts are known as dry—for example, the thick, black, crumbly extract of aloe—soft, or liquid.

Tinctures are obtained by maceration and filtering and are known as alcoholic if the macerating liquid is water and alcohol, or ethereal if it is ether. An alcoholic tincture added to a sugar solution is called an elixir. Ratafia is an unpalatable but beneficial elixir made with cherries. According to a recipe given in *Medicinal Plants* by Roberto Chiej (London, 1984), the elixir of life consists of thirty parts of aloe, five of rhubarb, five of gentian and five of aromatic calamus (or sweet sedge, *Acorus calamus*) dissolved in 1,000 parts of alcohol at 60°. It should be taken by the teaspoonful as an aromatic tonic.

Vegetal drugs macerated in wine, preferably white, produce medicinal wines. If they are distilled, the resulting liquids will only contain the volatile principles. Syrups are made by dissolving sugar in water over a low heat, allowing most of the water to evaporate and then mixing the residue with the therapeutic substances.

In the past medicinal oils were used for therapeutic massage to relieve aches and pains and were obtained by dissolving an oily extract in olive oil, or by boiling the drug in the oil and then filtering it. Ointments and unguents were made by mixing powdered drugs with fat, or by melting the fat with medicinal substances and allowing it to solidify. It is still possible to find old apothecaries' jars with labels such as 'unguent of poplar' or 'unguent of lily root'.

Essences, or essential oils, are extracted from certain plants without the application of heat—in the case of citrus fruits by pressing the peel. A second method of extraction is known as 'hot enfleurage'. This involves the use of 'fixed' solvents, such as ox or pork fat or refined olive oil, which do not evaporate at a normal temperature. The essence-producing plants are placed in copper containers in which the solvent has been heated to 40°C. When the plants have released their essential oils, they are replaced with fresh ones, until the fat is saturated.

An alembic (from the Arabic *al-anbiq*, meaning vase) is used to extract essential oils by distillation. It consists of three parts: the boiler, made of copper or glass, under which the source of heat is placed; the cap, or alembic proper; and the condenser. The plants to be used should be freshly picked at the height of their oil production and should not be laid on top of one another, to avoid fermentation. They are placed in baskets in the boiler, over water which is turned into steam by the heat. The steam causes the essential oils to be released. The vapour rises into the cap, where it expands and cools slightly. It then passes through the condenser, a spiral tube immersed in cold running water, which transforms the vapour back into liquid. This collects in a receiver, a sort of jug known as a Florentine flask, in which the essences and the water separate due to their different specific weights. The essential oils may then be decanted.

As an example of a treatment for a headache caused by indigestion, here is the recipe for melissa water which is easily prepared by one of the above methods. (*Melissa officinalis*, or lemon balm, belongs to the Labiatae, has pale green to yellowish, slightly hairy leaves and pinkish-white flowers.) Mix together twenty grams each of lemon balm and lemon peel, ten of ground nutmeg, five each of cloves, calamus root, cinnamon, angelica root and coriander seeds and 600 grams of alcohol at 95° and leave to macerate for ten days. Filter off the liquid, leave for a day and filter again. A few drops may be taken in a little water.

The perfume of Egypt

In 1760 Frederick V of Denmark organized a scientific expedition to Egypt and Arabia. Members of the expedition included the Swedish naturalist Peter Forskål, a pupil of Linnaeus, and Carsten Niebuhr, an engineer from Hanover. The scientists landed at Alexandria in September 1761 and the following month left for the Yemen. Linnaeus thought highly of his pupil Forskål, and predicted that he would make a great many discoveries. After landing at Alexandria, Forskål listed some 2,000 new species, but died of the plague in Arabia in 1763. His *Flora Aegyptiaco-Arabica* was later published by Niebuhr, who had seen his colleagues die one by one and was the only survivor of the expedition.

History has it that General Bonaparte had the volumes of Niebuhr's account on his desk when he summoned the scientists who were to accompany him on his conquest of Egypt, and whose reports led to the rediscovery of this ancient and mysterious land. Some of the scientists had been selected by Constantin François de Chasseboeuf, Count of Volney, philosopher and follower of Diderot, who had used a small legacy to undertake a four-year journey to Syria and Egypt (1782–86). He studied Arabic in a Maronite convent in Lebanon and travelled by camel; his *Voyage en Egypte et en Syrie* (Journey to Egypt and Syria; 1787) and above all his *Les Ruines ou Méditations sur les Révolutions des Empires* (Ruins, or Meditations on the Vicissitudes of Empires; 1791) met with great success.

The temple at Edfu, the ancient Apollinopolis Magna, was still buried under a village on the left bank of the Nile, some sixty miles upstream from Thebes (Luxor). Like many other ancient monuments in Egypt, it was excavated by the Frenchman Auguste Edouard Mariette during the latter half of the 19th century. It is a Ptolemaic temple, founded, according to the inscriptions, on 23 August of the year 237 BC by Ptolemy III Euergetes. It is interesting that the walls of a small room off the second colonnaded hall of the temple bear inscriptions consisting of long recipes for perfumes, including one for cyphi.

Plutarch, the Greek historian who lived during the 1st and 2nd centuries AD, observed in his *On Isis and Osiris* that perfumes were burnt every day by the Egyptians, who always laid 'the very greatest stress upon those practices which are conducive to health . . . Since, then, the air has not always the same consistency and composition, but in the night-time becomes dense and oppresses the body and brings the soul into depression and solicitude, as if it had become befogged and heavy, therefore, immediately upon arising, they burn resin on their altars, revivifying and purifying the air by its dissemination . . . Again at midday, when they perceive that the sun is forcibly attracting a copious and heavy exhalation from the earth and is combining this with the air, they burn myrrh on the altars; for the heat dissolves and scatters the murky and turgid concretions in the surrounding atmosphere.' Cyphi was prepared from sixteen ingredients (honey, wine, raisins, cyperus, resin, myrrh, ebony, seseli, lentiscus, bitumen from Judea, henbane, juniper, cardamom, cinnamon, calamus and terebinth), according to a formula found in the 'sacred writings', and it produced marvellous results. It softened the air, relaxed the body, induced sleep, made daily troubles recede and heightened the imagination so that it resembled 'a mirror, and makes it clearer no less effectively than did the notes of the lyre which the Pythagoreans used to employ before sleeping'.

An Homeric medicine

We read in the *Odyssey* that Telemachus, Odysseus' son, set sail in a desperate attempt to obtain news of his father who had been absent for twenty years, and landed on the Peloponnese. In Sparta, he was the guest of Menelaus, and was brought to tears by tales of the past and of his father. Then Helen added a drug to the sweet wine they were drinking which was effective against tears or wrath and conducive to a comfortable forgetfulness of worry or sorrow. In other words, she gave him what we would call a tranquillizer, which is later called *nepenthes*.

Over the centuries an interesting discussion arose as to which drug this was; the controversy was summarized at the beginning of the 18th century by Johann Albert Fabricius in his *Bibliotheca Graecae* (Hamburg, 1705), but new suggestions continued to be made. The following drugs were mentioned in the course of the controversy: gelatophyllis, hestiateris, oenothera, crocus, bugloss, lotus, Egyptian helenium, mandrake, Indian hemp, henbane, datura, opium. Some of them are referred to by Pliny in his *Historia Naturalis*, a survey of contemporary botanical knowledge written some six centuries after Homer.

According to Pliny, drinking gelatophyllis with wine and myrrh causes such uncontrollable laughter that palm wine with pine-kernels, pepper and honey had to be drunk in order to stop it. Hestiateris was another laughing drug used by the Persians. The therapeutic tradition associated with oenothera begins with Krateuas, physician to Mithridates the Great and already mentioned as the pioneer of illustrated herbals. The effect of this drug when drunk with wine was to induce hilarity and, if given to wild animals, to tame them.

The *Regimen sanitatis seu Flos Medicinae Salerni* (Health Regulations or the Flower of Salernitan Medicine) is an exposition in verse of the medical precepts of the School of Salerno, active from the Middle Ages until its closure in 1811 by Joachim Murat, King of Naples. It mentions both the crocus and the bugloss for their cheering properties:

Confortare crocus dicitur laetificando
membraque defecta confortat, hepar reparando.
Crocus comestus pulchrum dat semper odorem,
Omnem foetorem tollit et pellit amorem.

(The crocus brings comfort and good cheer, revives weary limbs and restores the liver; when eaten it imparts a pleasant smell, destroys all foetid odours, and encourages love.) Crocus, in this context, is the saffron crocus. A decoction of bugloss, according to the School of Salerno, 'livens up all banquets'. It was prescribed by the astrologer and alchemist Arnaldo di Villanova, physician to Pope Clement V, to cure melancholy.

The effects of the lotus are known from Odysseus' stay in the land of the lotus-eaters (probably on the Libyan coast): 'Any of them who ate the honey-sweet fruit of the lotus was unwilling to take any message back, or to go away, but they wanted to stay there with the lotus-eating people, feeding on lotus, and forgetting the way home.' (*Odyssey*, IX, 94–97; translated by R. Lattimore).

According to classical tradition, 'erba elenea' or Egyptian helenium confers beauty upon the female face. When drunk with wine it restrains men from quarrelling and makes them immune to snake bites.

A great deal has been written about the magical properties of mandrake: it keeps evil spirits at bay, quietens storms and uncovers buried treasures. It is an aphrodisiac, as testified by the fact that the Emperor Julian the Apostate called Venus 'mandragorotis', and the English accused Joan of Arc of keeping a root of mandrake underneath her cuirass. One of Boccaccio's stories tells how a famous surgeon called Mazzeo della Montagna had his patient drink a liquid distilled according to one of his own recipes before an operation, in order to keep him asleep for as long as the surgeon needed to cure him. Literary and medical research has identified Mazzeo della Montagna with a contemporary physician living in Salerno, Matteo Selvatico, who, quoting from the Arab physician Serapion wrote that surgeons give mandrake root to their patients when they have to cut or cauterize a limb.

Herodotus mentioned Indian hemp in connection with the Scythians: it was very similar to flax, and the Scythians threw the seeds onto red-hot stones; they burnt quickly, giving off more vapour than in a Greek tepidarium. This powerful sudorific had a peculiar effect on the Scythians, who would shout with happiness.

Henbane, *Hyoscyamus niger*, is a poisonous member of the Solanaceae family, which was used as an external analgesic. It can produce inebriation, as testified by Herman Boerhaave who was thus affected while preparing a soothing ointment.

Datura (jimson weed or thorn apple) is another poisonous solanaceous plant with narcotic properties. It is mentioned by the Portuguese physician Garcia da Orta, who spent most of his life in the Portuguese colony of Gôa, in his *Coloquios dos Simples, e Drogas he Cousas Mediçinais da India* (Discourses upon the Simples, and Drugs and Medicaments from India; 1563) which he published there: 'When thieves want to rob people, they mix these flowers with the food they are eating, with the result that all those who partake of the food are overcome by laughter and prodigality and of their own free will allow the thieves to rob them.'

Some fifty years ago the psychiatrist Lorenzo Gualino wrote a delightful essay summing up the controversy about Helen's drug. Telemachus finally fell into a refreshing sleep which Pliny believed was simply due to the wine, while others put it down to his being mesmerized by the charms of the Queen of Sparta. However, it seems likely that the mysterious *nepenthes* was an opiate—one of those secret Egyptian drugs which Helen, according to Homer, learnt about from the Egyptian queen, Polydamna.

Physicians and fashions

In the mid-18th century, Carlo Goldoni wrote a comedy entitled *La finta ammalata* (The Imaginary Valetudinary). The protagonist, as he tells us in his *Memoirs*, was based on a Mrs Medebac—'an excellent actress, strongly attached to her profession, but she was subject to fits of *ennui*; she was often ill, often imagined herself so, and sometimes nothing ailed her but her fits, which she had at her command'. In the play a consultation is staged: three physicians and the family surgeon gather by the bedside of the would-be invalid, who in fact is only in love with her doctor. Dr Onesti, the physician in charge, who knows the patient, concludes from the symptoms that a spiritual rather than a physical disturbance is the cause of her illness. Dr Buonatesta, having examined the patient, disagrees. Dr Malfatti agrees partly with the one, partly with the other. And the surgeon, having asked permission to speak, recommends a blood-letting. 'I am a doctor's son', wrote Goldoni, 'and have been a doctor myself for a while, and I blame equally those who praise and those who condemn medicine in general. When I had to write about this art, which one must perforce respect, I introduced three doctors into my play: one is prudent and honest, one is a quack, the third is ignorant; they represent the three types of doctors one can come across. God protect us from the last mentioned, but the second type is even more dangerous.' Having outlined the plot, he continues: 'The plot is very simple, but the play was nonetheless well received and applauded; its success may have been due to the skill of the actress, who enjoyed portraying herself and could do so without effort or restraint. The different characters of the three doctors and of a gossiping, deaf apothecary who misunderstood everything and preferred by far to read the paper than the prescriptions, also contributed to the play's success.'

Blood-letting, recommended by the family surgeon for the would-be invalid in Goldoni's play, was for long a universal remedy. Cavour, the Italian statesman, died at dawn, on 6 June 1861, which was a Thursday. On the previous Saturday he had been bled twice; on the Monday this was again attempted but no blood came out of his veins. On the evening of the Wednesday, cupping-glasses were applied to his neck and vesicatories to his legs. And yet medicine, as a science, moved forward.

In the 18th century, the colleagues of Christoph Jakob Trew included Leopold Auenbrugger, a physician who discovered the percussion method of diagnosing diseases. The son of an innkeeper, he was accustomed to seeing his father knocking on the wine casks with his fingers to find out how much wine was left, and he applied the same principle when examining patients. Another notable physician was Giovanni Battista Morgagni, described by Carlo Ottavio Castiglione as belonging to the small number of the truly great, not only in the field of medicine, but also in the history of science; yet another was Edward Jenner who discovered vaccination against smallpox. The public, however, was always responsive to fashionable novelties, and this brought Franz Mesmer to the fore.

Mesmer was born in Weil, a village on Lake Constance. He graduated from Vienna after having written a thesis on *De planetarum influxu*, the influence of planets on human organisms. He then developed his theory of animal magnetism and of the therapeutic effect of the transfer of magnetic flux through the laying on of hands. In short, he was a 'healer'.

In 1778, distrusted by his Viennese colleagues, Mesmer emigrated to Paris where, as a protégé of Queen Marie Antoinette, he enjoyed great success. He was so much in demand that it took several weeks to be admitted to his surgery. In a dimly lit room where the air was scented and soft music played in the background, patients were seated around a tub containing sulphuric acid diluted with water, from which iron spikes protruded. The magnetic flux was transmitted to them by the spikes, by holding hands and forming a human chain, and by the staff Mesmer held.

In 1794 an authoritative scientific commission was set up to look into the matter. Members included one of the de Jussieu family of French botanists, Benjamin Franklin and Antoine-Laurent Lavoisier; they concluded that there was no proof of the existence of animal magnetism: 'imagination without magnetism produced convulsions, and magnetism without imagination produced nothing'. Mesmer left Paris and died in obscurity in 1815 at the age of eighty-one, by Lake Constance. The therapeutic effectiveness of the 'imagination' had not yet been discovered.

The plant hunters

. . . at nine o'clock they came back on board carrying various plants, flowers, etcetera, most of them unknown in Europe: this was the reason for their great value.

James Cook, *Log-book*, January 1769

Pineapple

Pineapple

Ananas sativus, Bromeliaceae

This plant, now widely cultivated in all tropical areas, was known centuries ago to the Guaraní Indians of Paraguay, who introduced it to other areas. The oldest known drawing of it can be found in a manuscript sent to King Ferdinand of Spain by Oviedo y Valdez in 1513. Valdez had discovered the plant in Haiti, where it was under cultivation, and praised its delicious fruit, a great aid to digestion.

Monarda

Monarda, Labiatae

This genus of North American plants was named after the Seville botanist Nicolò Monardes, who was the first person to publish a catalogue of the plants of the area, which he had gathered and classified. Like most of the Labiatae, monardas are highly aromatic (they contain thymol) and were therefore used by the Indians to prepare diuretic teas or to add flavour to salads.

Monarda

Lycium

Lycium

Lycium, Solanaceae

Mattioli, the Sienese physician who translated and illustrated the work of Dioscorides, one of the most famous classical treatises on herbalism, dealt at length, in his *Discorsi*, with the origin of the name *lycium*, applied to plants which were native to the Roman provinces of Lycia or Cappadocia. Mattioli was rather doubtful as to the exact classification of the lycium, mainly because the name initially designated plants of the genus *Rhamnus*, and was later applied to these members of the Solanaceae family. The lycium has small, edible fruits resembling tomatoes.

Pontederia

Ceanothus

Pontederia

Pontederia, Pontederiaceae

Linnaeus named this genus after the director of the botanic garden in Padua, Giulio Pontedera (1688–1757). All the plants of the genus are native to tropical regions and favour swampy habitats, spreading rapidly. At one time the genus included the water lily.

Ceanothus

Ceanothus americanus, Rhamnaceae

These North American plants grow in sandy, rather arid habitats from Mexico to Montana. Two species, *Ceanothus americanus* and *C. azureus*, have strong historical connotations. *C. azureus* is known in America as 'liberty tea' because its leaves were widely used as a substitute after a cargo of real tea was unloaded into the sea at Boston during one of the incidents which sparked off the War of Independence. The leaves of *C. americanus* are called 'New Jersey tea' because they were brewed by soldiers during the American Civil War.

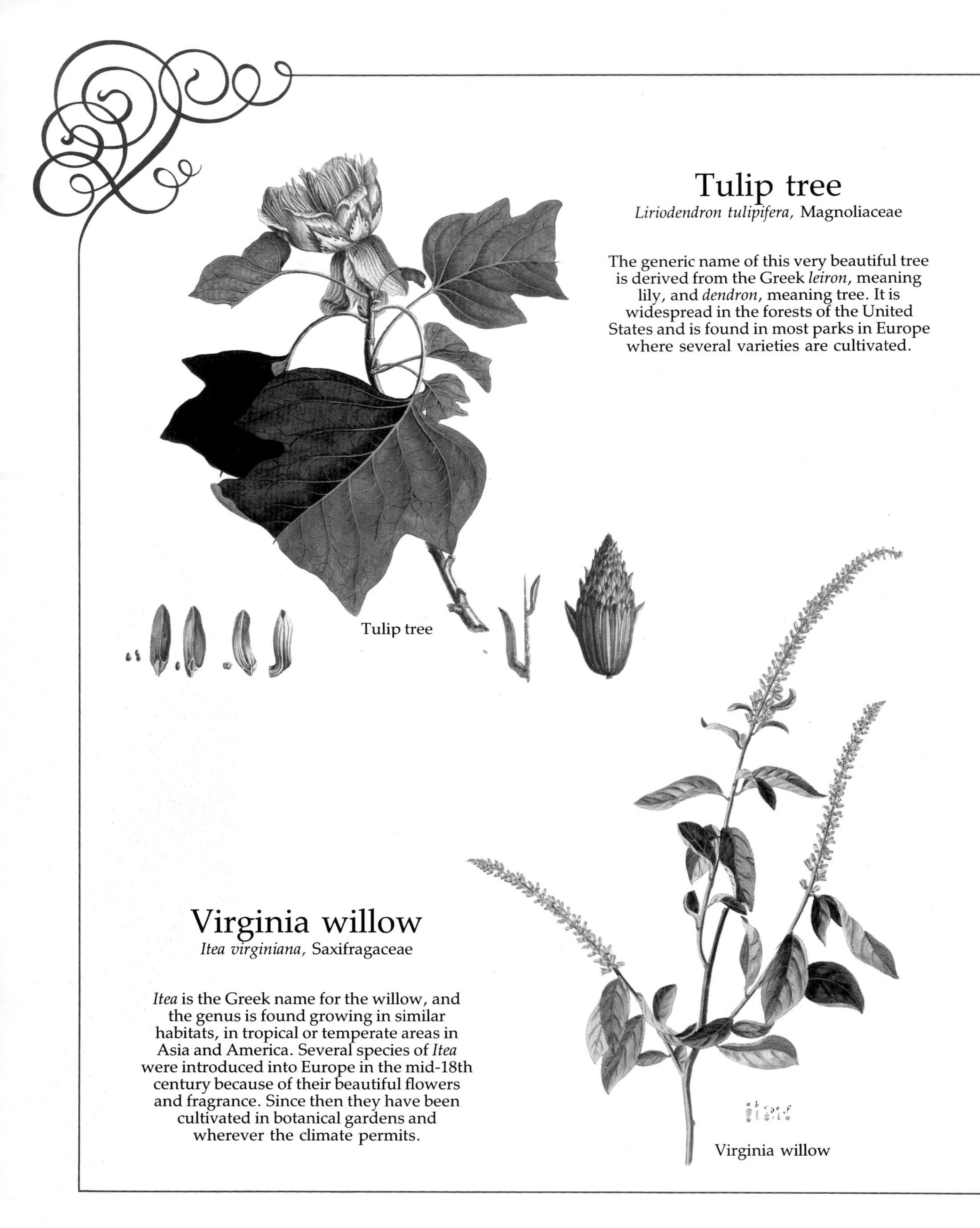

Tulip tree

Liriodendron tulipifera, Magnoliaceae

The generic name of this very beautiful tree is derived from the Greek *leiron*, meaning lily, and *dendron*, meaning tree. It is widespread in the forests of the United States and is found in most parks in Europe where several varieties are cultivated.

Tulip tree

Virginia willow

Itea virginiana, Saxifragaceae

Itea is the Greek name for the willow, and the genus is found growing in similar habitats, in tropical or temperate areas in Asia and America. Several species of *Itea* were introduced into Europe in the mid-18th century because of their beautiful flowers and fragrance. Since then they have been cultivated in botanical gardens and wherever the climate permits.

Virginia willow

Magnolia

Magnolia grandiflora, Magnoliaceae

Linnaeus named the genus after Pierre Magnol (1638–1715) who taught both botany and medicine at Montpellier University. In America the magnolia was widely used in popular medicine both as a febrifuge and in the manufacture of liqueurs.

Magnolia

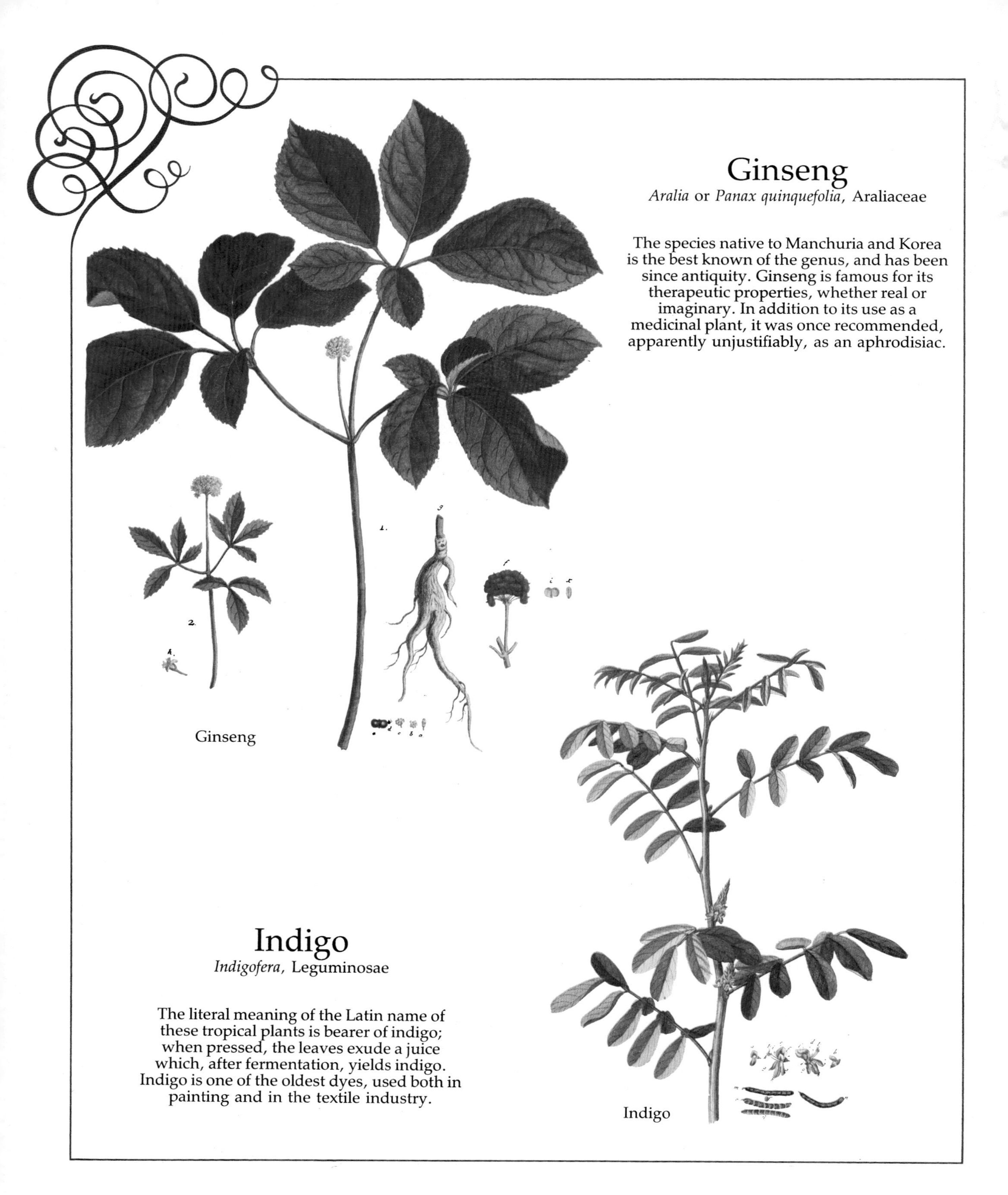

Ginseng

Aralia or *Panax quinquefolia*, Araliaceae

The species native to Manchuria and Korea is the best known of the genus, and has been since antiquity. Ginseng is famous for its therapeutic properties, whether real or imaginary. In addition to its use as a medicinal plant, it was once recommended, apparently unjustifiably, as an aphrodisiac.

Ginseng

Indigo

Indigofera, Leguminosae

The literal meaning of the Latin name of these tropical plants is bearer of indigo; when pressed, the leaves exude a juice which, after fermentation, yields indigo. Indigo is one of the oldest dyes, used both in painting and in the textile industry.

Indigo

Balsam poplar

Populus balsamifera, Salicaceae

One of the best known of the American species, it belongs to the Tacamahac group of poplars renowned for the balsam, an expectorant and stimulant, which can be extracted from the leaves.

Balsam poplar

Sensitive plant

Mimosa sensitiva, Mimosaceae

This and similar species such as *Mimosa pudica* have been familiar since the mid-17th century when they were first brought to Europe from Brazil. Their most interesting feature is their sensitivity to certain stimuli (touch or heat) which will cause their leaves to close and droop as if dead. This is due to the movement of the liquid contained in small groups of cells situated where the leaves and leaflets join the stem.

Sensitive plant

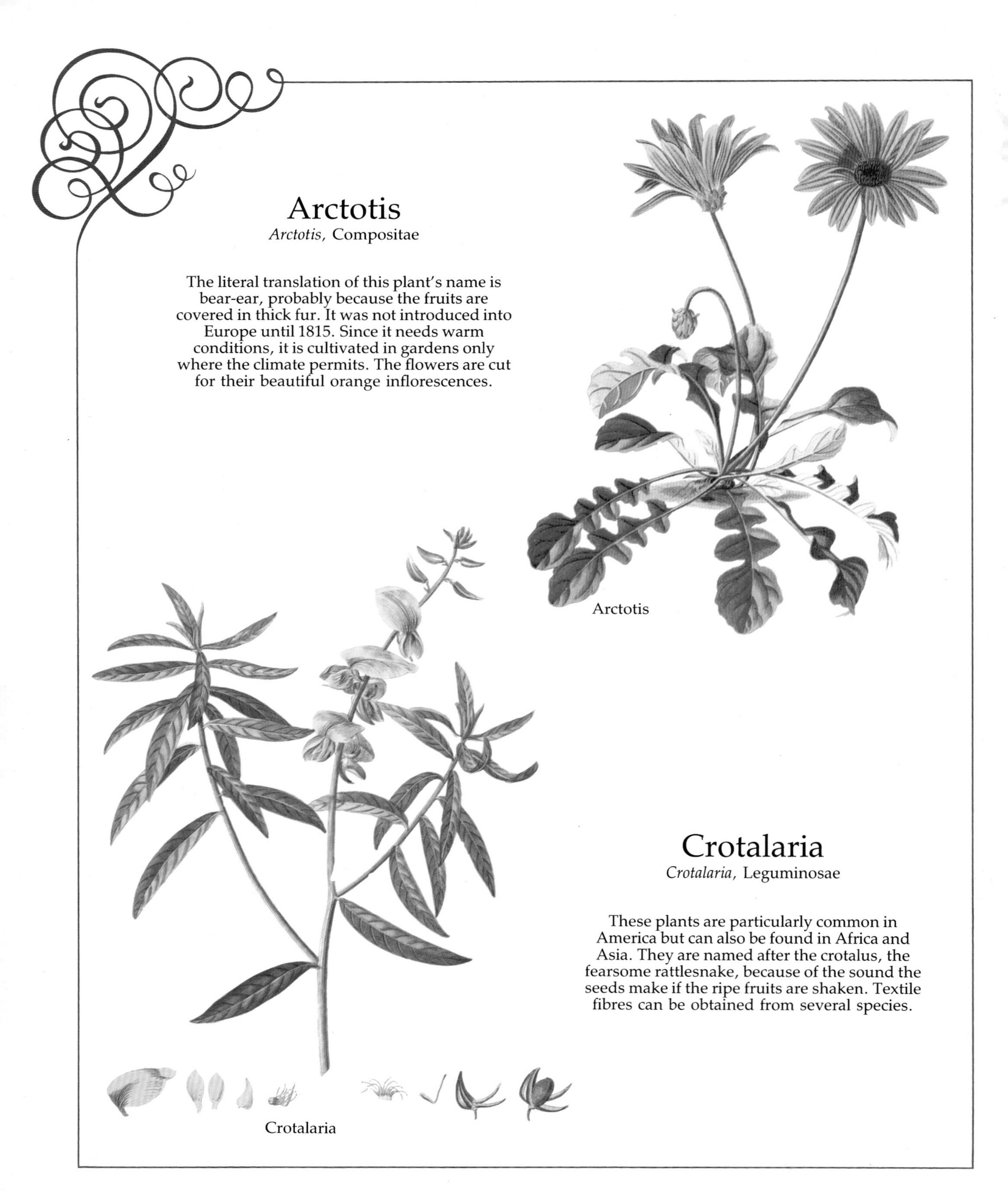

Arctotis

Arctotis, Compositae

The literal translation of this plant's name is bear-ear, probably because the fruits are covered in thick fur. It was not introduced into Europe until 1815. Since it needs warm conditions, it is cultivated in gardens only where the climate permits. The flowers are cut for their beautiful orange inflorescences.

Arctotis

Crotalaria

Crotalaria, Leguminosae

These plants are particularly common in America but can also be found in Africa and Asia. They are named after the crotalus, the fearsome rattlesnake, because of the sound the seeds make if the ripe fruits are shaken. Textile fibres can be obtained from several species.

Crotalaria

Cereus

Selenicereus grandiflorus, Cactaceae

The genus *Cereus*, according to early systems of classification, contained so many widely differing succulents that it has been divided into several distinct genera. *Selenicereus* is one of the most spectacular, with large flowers that open in the moonlight and are pollinated by a species of bat.

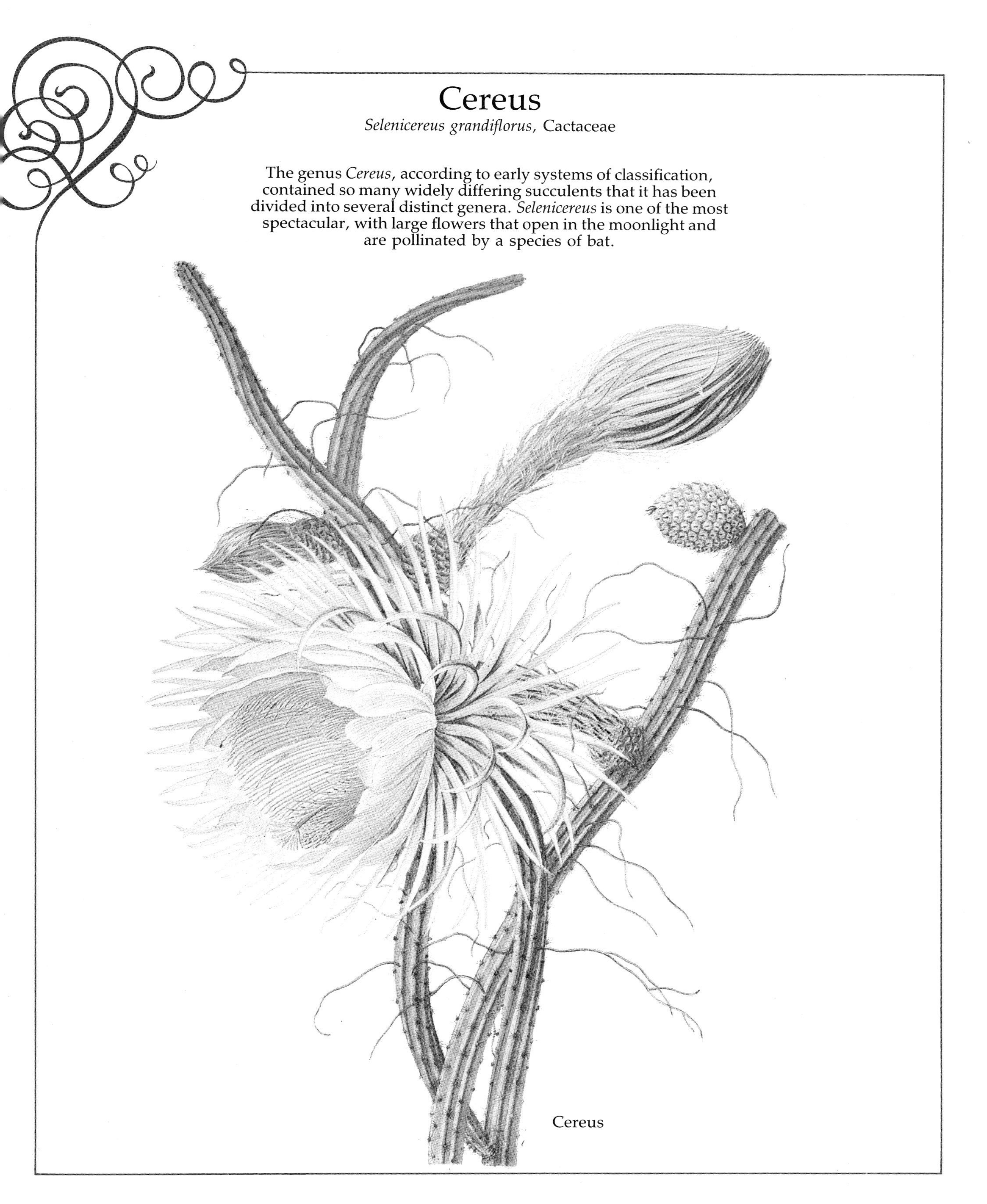

Cereus

Banana

Musa paradisiaca, Musaceae

The banana is probably native to the Far East, where wild species can still be found producing fruits with fertile seeds which ensure the preservation and diffusion of the species. The Arabs imported the cultivated variety with seedless fruits (parthenocarpic) into Africa. It is thought that the tree was introduced into America by the Spanish, who planted a few specimens in Santo Domingo, in 1516. From there the banana spread to the rest of the West Indies but did not reach South America until a little later, when it was introduced by the Portuguese.

Banana

Maple

Maple

Acer, Aceraceae

The genus was known to the Romans, who, according to some authors, chose the name *Acer* because of the sharp points of the lobed leaves. There are many American species, and most of the maples with composite leaves are not native to Europe. Many maples are now widely planted in towns and gardens for their ornamental colouring. They have characteristic winged fruits which, when they fall from the branch, can be carried far from the mother plant by the wind.

Crataegus

Crataegus tomentosa, Rosaceae

This North American shrub was initially classified as belonging to the genus *Mespilus* but rightly belongs to *Crataegus*. The best-known *Crataegus* species are the azarole (*C.azarolus*), cultivated for its edible fruits, and the hawthorn (*C.monogyna*). According to a French legend, the hawthorn moans every year on Good Friday, in memory of how it was (supposedly) used to make the crown of thorns for Jesus.

Crataegus

Bocconia

Bocconia

Bocconia racemosa, Papaveraceae

This is one of the few shrubby plants in the Papaveraceae family. It was named *Bocconia* in honour of Paolo Bocconi, a 17th-century Sicilian botanist who joined the Cistercian order, taking the name of Silvius.

Belladonna lily

Amaryllis, Amaryllidaceae

This genus used to be known by the old name of *Lilio-narcissus* because some species were originally classified with species of narcissi. *Lilio* is a reference to the beauty of these plants, which are native to tropical America and were probably introduced into Europe in the 17th century. The current generic name, *Amaryllis,* is also a reference to the beauty of the flowers; it is derived from the Greek *amarusso,* meaning to be resplendent.

Belladonna lily

Ixia

Ixia

Ixia, Iridaceae

Ixias are widely used by florists as their blooms last a long time even when cut. The first plants to be imported from Africa in the 18th century were grown in the Jardin des Plantes in Paris. The name, from the Greek word for a sticky substance, refers to the rubbery, sticky sap exuded by the cut bulbs.

Hymenocallis

Hymenocallis, Amaryllidaceae

This genus was separated from the very similar *Pancratium* only at the beginning of the 19th century, hence it is classified under the latter in earlier texts. All the species of *Hymenocallis* are of American origin, except for one which was also found in Africa. Their name refers to the 'beautiful membrane' which joins the lower part of the stamens. *Pancratium* species, two of which are native to the Mediterranean, differ only in the number of seeds contained in the fruits.

Hymenocallis

Petiveria

Petiveria, Phytolaccaceae

This genus was named after James Petiver, a botanist and pharmacist who lived in London at the turn of the 18th century. The plants are native to the tropical and subtropical areas of America, from Texas to Paraguay. As medicinal plants, they are used in poultices to relieve contusions.

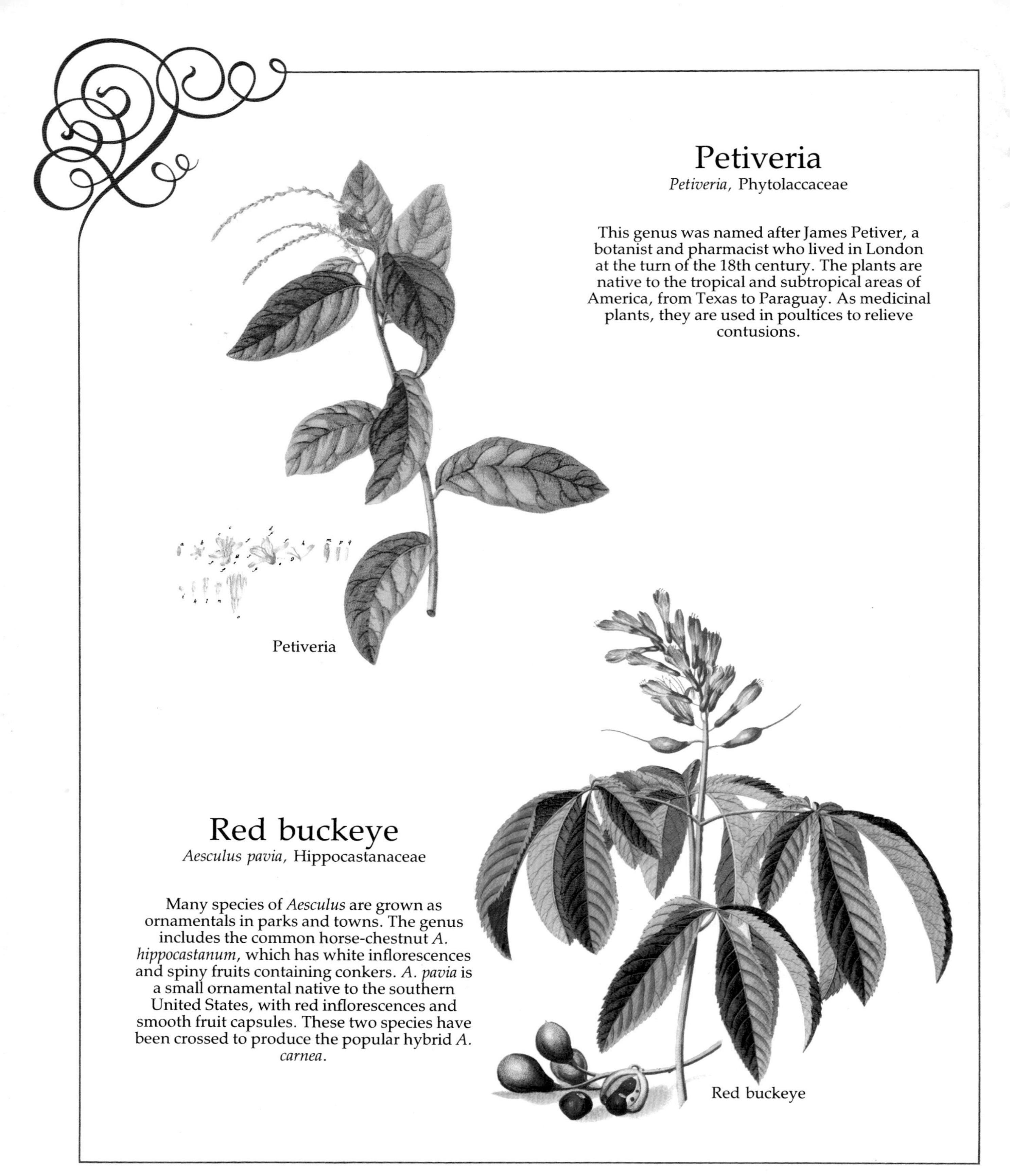

Petiveria

Red buckeye

Aesculus pavia, Hippocastanaceae

Many species of *Aesculus* are grown as ornamentals in parks and towns. The genus includes the common horse-chestnut *A. hippocastanum,* which has white inflorescences and spiny fruits containing conkers. *A. pavia* is a small ornamental native to the southern United States, with red inflorescences and smooth fruit capsules. These two species have been crossed to produce the popular hybrid *A. carnea*.

Red buckeye

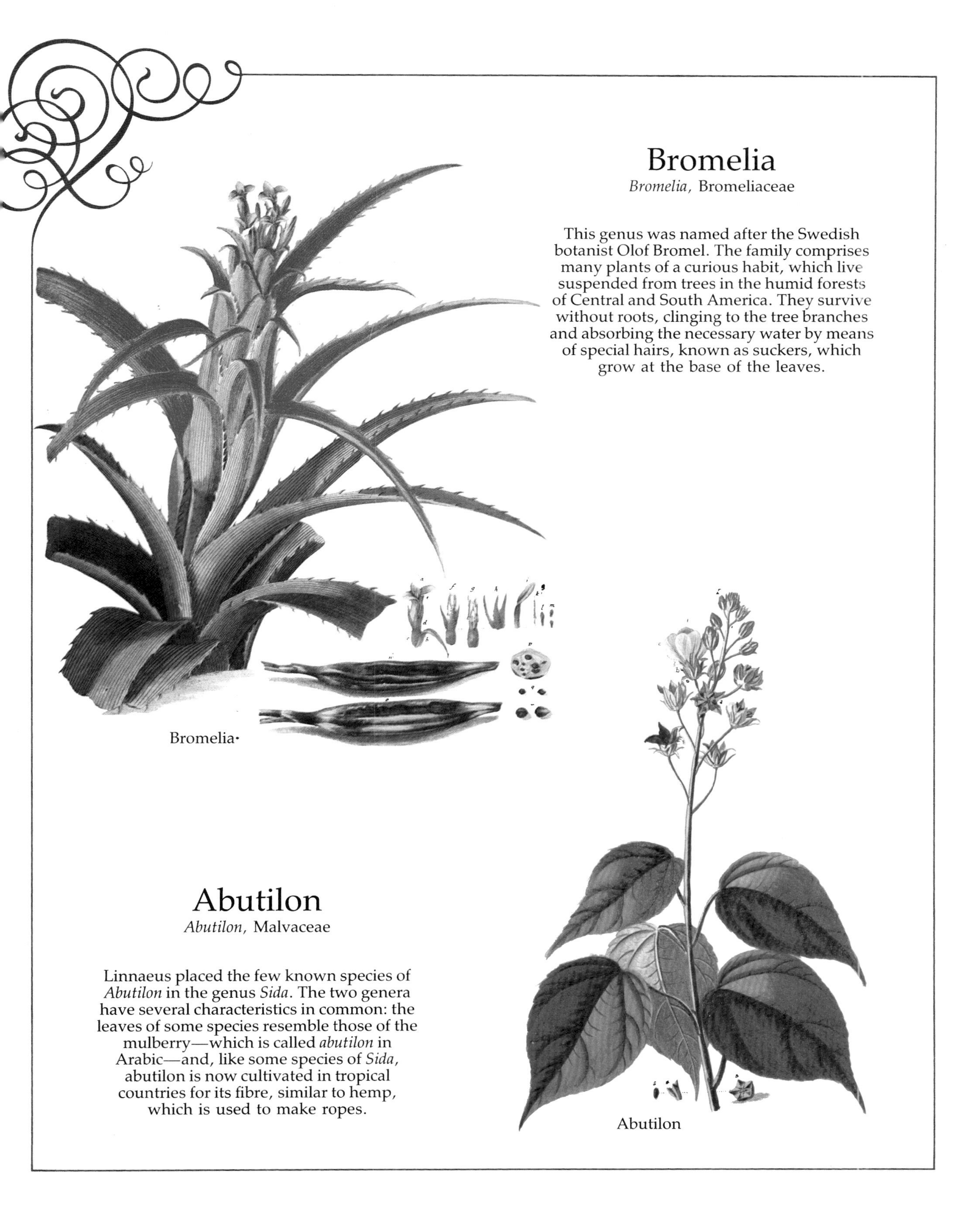

Bromelia

Bromelia, Bromeliaceae

This genus was named after the Swedish botanist Olof Bromel. The family comprises many plants of a curious habit, which live suspended from trees in the humid forests of Central and South America. They survive without roots, clinging to the tree branches and absorbing the necessary water by means of special hairs, known as suckers, which grow at the base of the leaves.

Bromelia

Abutilon

Abutilon, Malvaceae

Linnaeus placed the few known species of *Abutilon* in the genus *Sida*. The two genera have several characteristics in common: the leaves of some species resemble those of the mulberry—which is called *abutilon* in Arabic—and, like some species of *Sida*, abutilon is now cultivated in tropical countries for its fibre, similar to hemp, which is used to make ropes.

Abutilon

Fig

Ficus, Moraceae

About 2,000 species are included in this genus. Some yield a latex similar to rubber when their stem is incised; some American species were used to build suspension bridges; others are liana-like (climbing and twining) in habit and are termed 'strangler' figs as they kill the trees they cling to.

May apple

Podophyllum peltatum, Barberidaceae

This species is widespread in the south-eastern areas of the United States. The generic name, from the Greek *podos*, meaning foot, and *phyllon*, meaning leaf, derives from the foot-like shape of the leaves. The yellow fruits are edible, with a bitter-sweet taste. The other parts of the plant contain laxative principles.

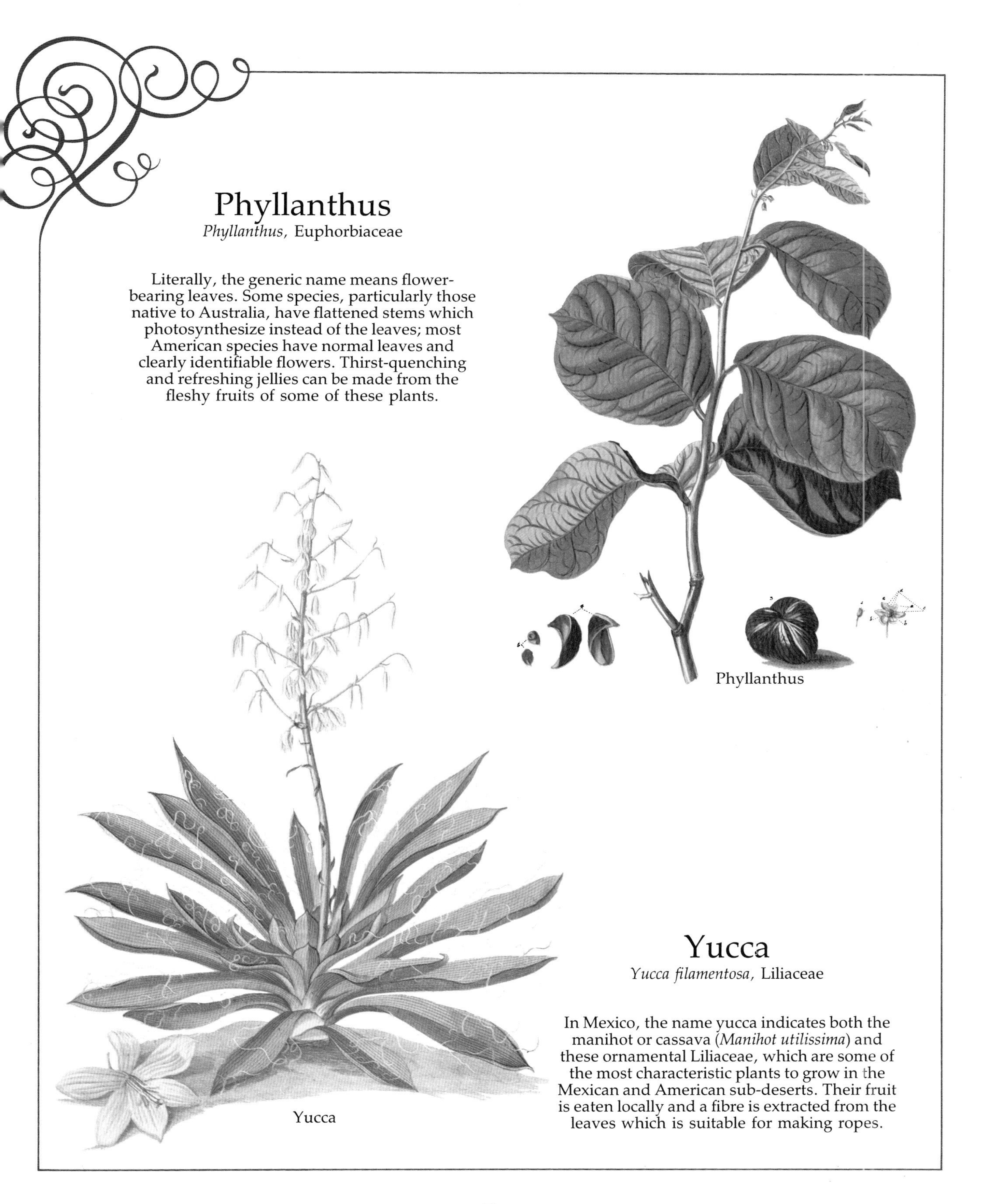

Phyllanthus

Phyllanthus, Euphorbiaceae

Literally, the generic name means flower-bearing leaves. Some species, particularly those native to Australia, have flattened stems which photosynthesize instead of the leaves; most American species have normal leaves and clearly identifiable flowers. Thirst-quenching and refreshing jellies can be made from the fleshy fruits of some of these plants.

Phyllanthus

Yucca

Yucca

Yucca filamentosa, Liliaceae

In Mexico, the name yucca indicates both the manihot or cassava (*Manihot utilissima*) and these ornamental Liliaceae, which are some of the most characteristic plants to grow in the Mexican and American sub-deserts. Their fruit is eaten locally and a fibre is extracted from the leaves which is suitable for making ropes.

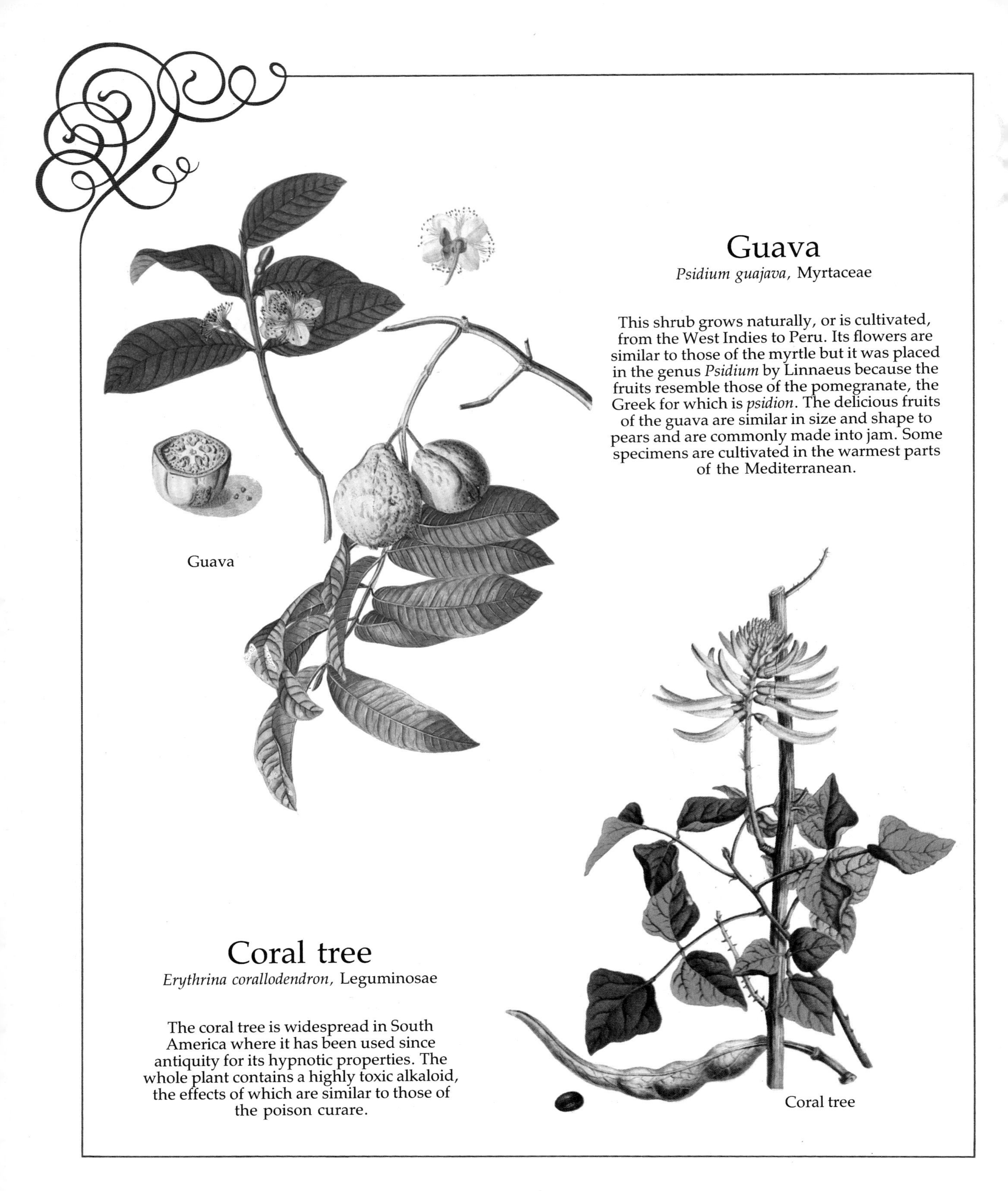

Guava

Psidium guajava, Myrtaceae

This shrub grows naturally, or is cultivated, from the West Indies to Peru. Its flowers are similar to those of the myrtle but it was placed in the genus *Psidium* by Linnaeus because the fruits resemble those of the pomegranate, the Greek for which is *psidion*. The delicious fruits of the guava are similar in size and shape to pears and are commonly made into jam. Some specimens are cultivated in the warmest parts of the Mediterranean.

Guava

Coral tree

Erythrina corallodendron, Leguminosae

The coral tree is widespread in South America where it has been used since antiquity for its hypnotic properties. The whole plant contains a highly toxic alkaloid, the effects of which are similar to those of the poison curare.

Coral tree

Papaya

Carica papaya, Caricaceae

A shrub widely cultivated in the tropics, it is probably a native of the West Indies. Vasco da Gama thought the fruits so delicious that he described the papaya as 'the golden tree of eternal youth'. In the 16th century it was described as a melon containing a medicinal latex. The fruits contain papain, a protein-digesting enzyme which is an aid to digestion.

Papaya

Blood lily

Haemanthus, Amaryllidaceae

These bulbous plants were first introduced into France in 1629 from the area around the Cape of Good Hope. The botanist Joseph Pitton de Tournefort placed them in this genus, the name of which is derived from the Greek word for blood. The plants have very unusual inflorescences and are cultivated as ornamentals all over the world.

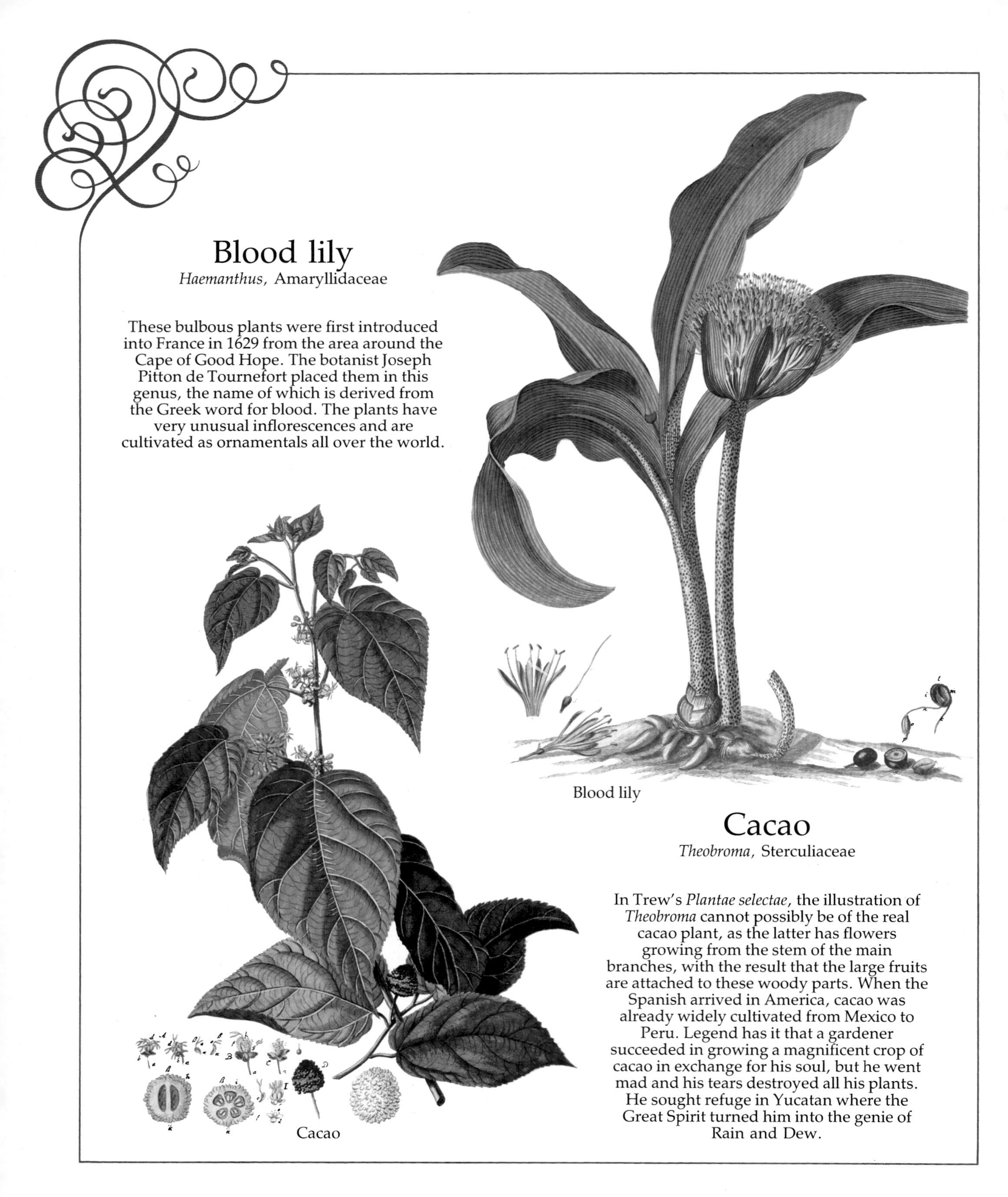

Blood lily

Cacao

Theobroma, Sterculiaceae

In Trew's *Plantae selectae*, the illustration of *Theobroma* cannot possibly be of the real cacao plant, as the latter has flowers growing from the stem of the main branches, with the result that the large fruits are attached to these woody parts. When the Spanish arrived in America, cacao was already widely cultivated from Mexico to Peru. Legend has it that a gardener succeeded in growing a magnificent crop of cacao in exchange for his soul, but he went mad and his tears destroyed all his plants. He sought refuge in Yucatan where the Great Spirit turned him into the genie of Rain and Dew.

Cacao

Sida

Sida, Malvaceae

It is something of a mystery as to why Linnaeus should have called these plants *Sida*, since Theophrastus had used this name for the white water lily which has nothing in common with these Malvaceae. Their stamens are joined into a single central column and the fruits grow together in a cluster. Some species yield a fibre used in rope-making in tropical countries.

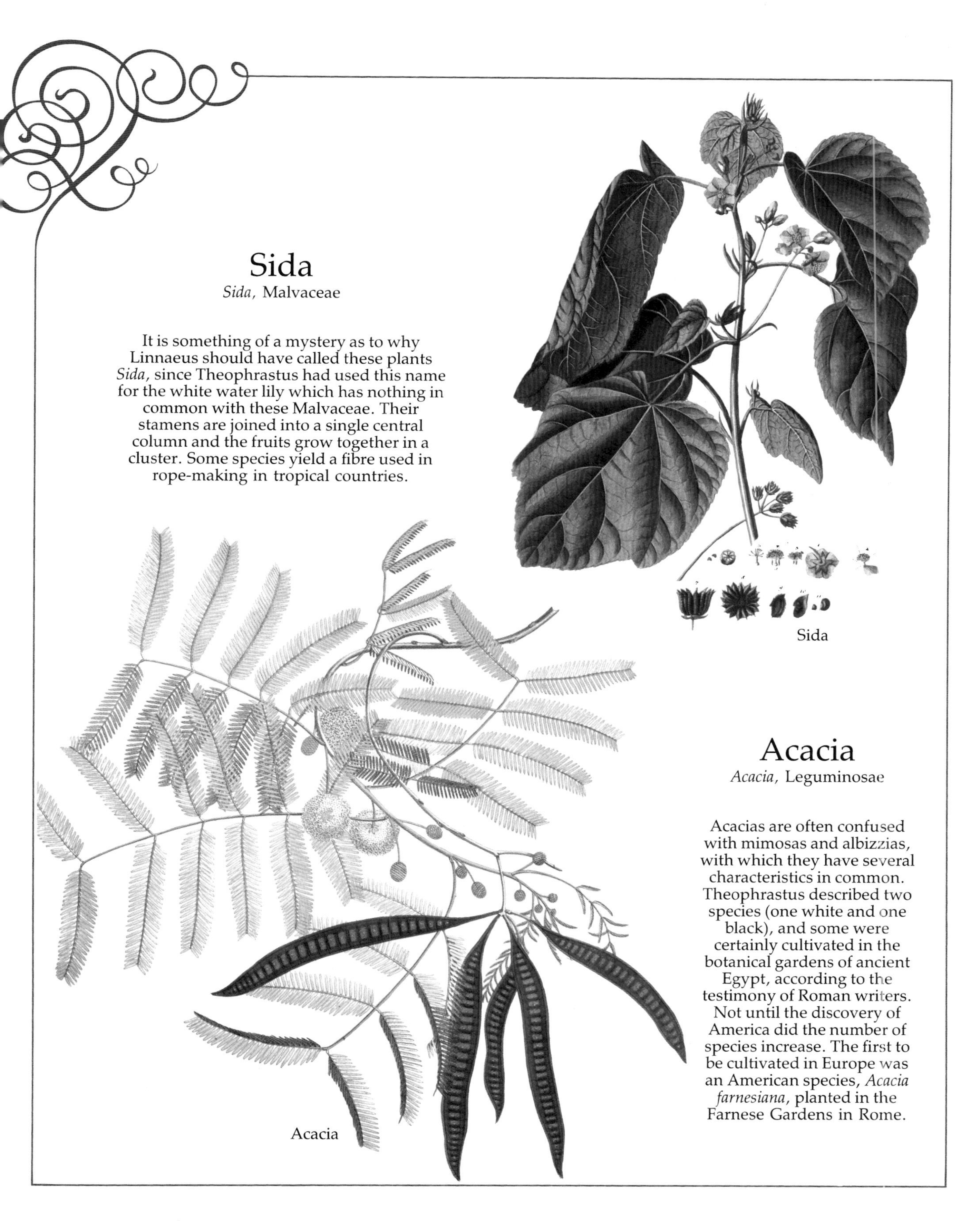

Sida

Acacia

Acacia, Leguminosae

Acacias are often confused with mimosas and albizzias, with which they have several characteristics in common. Theophrastus described two species (one white and one black), and some were certainly cultivated in the botanical gardens of ancient Egypt, according to the testimony of Roman writers. Not until the discovery of America did the number of species increase. The first to be cultivated in Europe was an American species, *Acacia farnesiana*, planted in the Farnese Gardens in Rome.

Acacia

Sandbox tree

Hura crepitans, Euphorbiaceae

Linnaeus used the local name for this genus, native to tropical North America. The specific name refers to the fact that the ripe fruits, which grow to the size of oranges, explode with a sharp bang, scattering thousands of tiny seeds. These closely resemble the sand in an hourglass—hence the vernacular name of sandbox tree.

Sandbox tree

Sweetsop, soursop

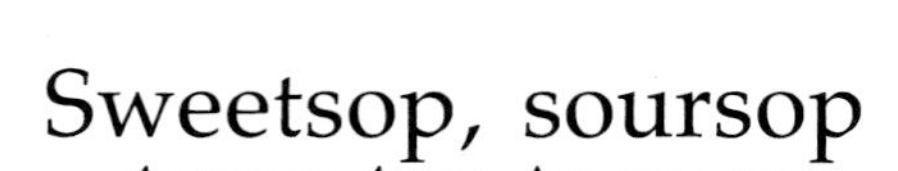

Sweetsop, soursop

Annona or *Anona*, Annonaceae

Various species of *Annona* (*A.squamosa, A.muricata*), mostly natives of tropical America, have long been cultivated in warm areas for their fleshy fruits. The name originally given them by Linnaeus is sometimes spelt with only one 'n', after the vernacular name *anon* by which the plants are known in Haiti.

Frangipani

Plumeria, Apocynaceae

About fifty species are included in this genus.
Most of them have highly scented flowers.
They are common from the West Indies to Peru
and are named after Charles Plumier, a French
explorer of the 17th century.

Frangipani

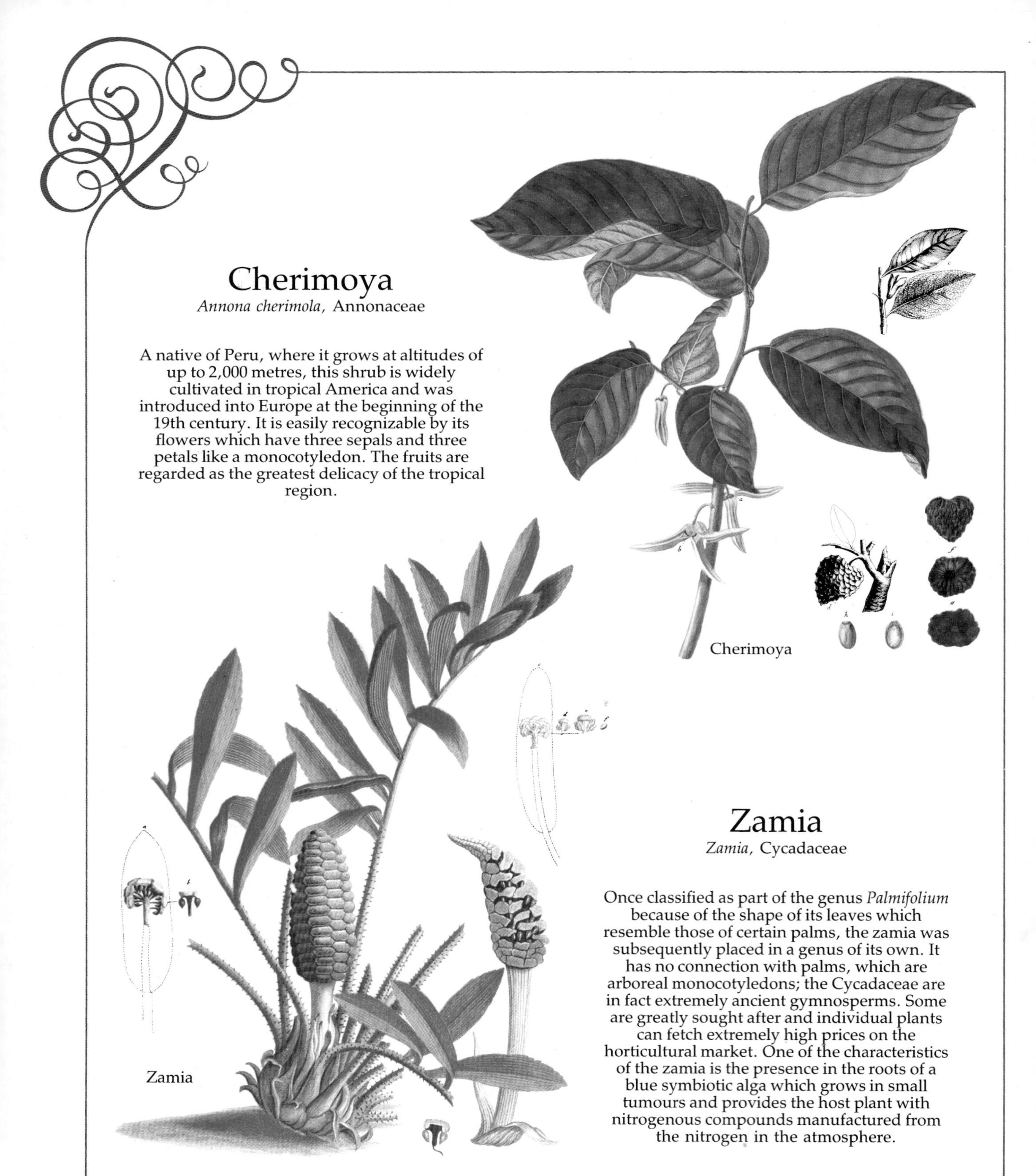

Cherimoya

Annona cherimola, Annonaceae

A native of Peru, where it grows at altitudes of up to 2,000 metres, this shrub is widely cultivated in tropical America and was introduced into Europe at the beginning of the 19th century. It is easily recognizable by its flowers which have three sepals and three petals like a monocotyledon. The fruits are regarded as the greatest delicacy of the tropical region.

Zamia

Zamia, Cycadaceae

Once classified as part of the genus *Palmifolium* because of the shape of its leaves which resemble those of certain palms, the zamia was subsequently placed in a genus of its own. It has no connection with palms, which are arboreal monocotyledons; the Cycadaceae are in fact extremely ancient gymnosperms. Some are greatly sought after and individual plants can fetch extremely high prices on the horticultural market. One of the characteristics of the zamia is the presence in the roots of a blue symbiotic alga which grows in small tumours and provides the host plant with nitrogenous compounds manufactured from the nitrogen in the atmosphere.

The plant hunters

Columbus's discoveries

As Christopher Columbus discovered on first approaching Cuba, the tropics gave off a marvellous scent of flowers. That characteristic, exotic fragrance of lush vegetation which wafts over the sea with the breeze must have made a lasting impression on those early sailors who had spent weeks on board foul-smelling ships. We are told about Columbus's reaction by his natural son, Ferdinand, in his *Historie della vita e dei fatti di Cristoforo Colombo* (The life and deeds of Christopher Columbus). Ferdinand was born in 1488 in Cordoba to a Spanish mother, Doña Maria de Toledo. His father was in Cordoba trying to raise money and sponsorship for his visionary plans. Although not interested in botany in a scientific sense, Christopher Columbus was the first person to take plants from the Old World across the Atlantic, and to bring American or, as they were considered at the time, Indian specimens back to Europe.

In the case of tobacco the person whose name has become associated with it is Jean Nicot, lord of Villemain—hence the word nicotine. However, this French diplomat merely introduced tobacco to France from Portugal, where he was ambassador from 1559–62, by sending tobacco powder to Queen Catherine de' Medici because it was said to be a cure for headaches. Responsibility for its discovery lies elsewhere. A note in Columbus's *Journal* (translated by Cecil Jane, 1968) dated Monday, 15 October 1492, three days after he first landed on the other side of the Atlantic, refers to tobacco leaves: 'Being in the middle of the channel between these two islands, that of Santa Maria and this large island, to which I gave the name of Fernandina [Rum Cay and Long Island in the Bahamas], I found a man alone in a canoe on his way from the island of Santa Maria to that of Fernandina. He was carrying with him a piece of bread, about as large as a fist, and a gourd of water and a piece of brown earth, powdered and then kneaded, and some dried leaves, which must be a thing highly prized among them, since already at San Salvador they presented me with some of them.'

Some days later, on 6 November, Columbus was off the Cuban coast and sent a couple of men to explore the interior: 'On the way the two Christians found many people, who were on their way to their villages, men and women, with a brand in their hands and the herbs for smoking, which they are in the habit of using.'

There is an illustration of *el tabaco* in a treatise by a physician from Seville, dealing with 'everything relating to our West Indies which is useful to the practice of medicine'. He says that both Negroes and Indians, tired out by dancing, were quite restored by smoking tobacco and were able to resume work afresh. 'They derive so much pleasure [from smoking] that they always want to do it, even when they are not tired,' so much so that their owners 'punish them and have the tobacco burnt, so that they cannot smoke it'. Agnes Arber, an expert on ancient herbals, suggests that this physician from Seville, who never crossed the Atlantic, may have confused tobacco with coca. In Ferdinand Columbus's account of his father's fourth journey, on which he himself went, aged fourteen, he is also probably referring to coca when he writes of the people of Veragua (Panama): 'The cacique and the chieftains never ceased to chew a dry herb, and sometimes added a certain powder which they carried with the herb, and this does not seem to me a good thing at all.'

In his version of the reconnaissance trip to the Cuban hinterland which is recorded in Columbus's *Journal* for 6 November 1492, Ferdinand also mentions potatoes—'cooked roots, resembling chestnuts in flavour'—and maize—'another grain which they call *mahiz*, which tastes excellent whether boiled or roasted or ground into flour and cooked'. The pineapple makes its first appearance during the second journey, to Guadeloupe: '. . . fruits which look like our pine cones, but much bigger; they are full of thick flesh like that of a melon, but of a much more delicate scent and flavour; these fruits grow all over the countryside on plants similar to lilies or aloes [but] the cultivated ones are even better, as we later found out.'

It was during the fourth journey that they met 'a canoe as long as a galley and eight feet wide, carved out of a single piece of wood, which was sailing, loaded with goods, from the west towards New Spain'. (New Spain was Mexico, which in the years between Columbus's fourth journey and the publication of Ferdinand's *Historie*, had been conquered by the Spaniards.) Columbus's sailors seized the goods in the canoe and found that they included 'several of those almonds which the people of New Spain use as coins, and which they seem to prize very highly, because I noticed that when they were taken aboard with all their things, some of these almonds were dropped and they all immediately bent down to pick them up, almost as if they had dropped one of their own eyes'. The canoe was seized off the Yucatan coast and the 'almonds' were cacao beans.

Montezuma's gardens

The gardens of the Aztec Emperor at Tenochtitlán were certainly as splendid as everything else in the great capital. They contained flowers and aromatic and medicinal plants native to the Aztecs' vast dominions, but they were destroyed by the Spanish in 1521. In 1552 a local, self-taught physician who was named Martin de la Cruz by the Spaniards, wrote a herbal of Mexican flora in the Nahuatl language. Another Aztec, Juan Badonio, translated it into Latin. Unfortunately, the manuscript (*Codex Barberini Lat.241*) remained buried in the vaults of the Vatican Library until some fifty years ago. Twenty years after it had been written and translated by the two Mexicans, Philip II sent his physician, Francisco Hernandez, to explore the natural resources of New Spain (Mexico). Hernandez remained there from 1571 to 1577 and brought back the manuscripts and illustrations for his *Rerum Medicarum Novae Hispaniae Thesaurum* (Thesaurus of New Spain's medical knowledge), which was bound in sixteen volumes, although its author did not have the satisfaction of seeing it printed. The manuscripts were destroyed in a fire in the Escorial; fortunately, the King had had extracts taken from them which were published in various editions during the 17th century. The medical and botanical knowledge of the Aztecs thus remains an academic curiosity.

Captain Bligh's breadfruit trees

Captain Bligh transported these trees from Tahiti to the botanical gardens of St Vincent in the West Indies in 1793. He had tried to do so before, but had been prevented by the famous mutiny which took place in 1789 on his ship, the *Bounty*. In 1736 manioc, the euphorbiaceous plant which yields cassava flour, or tapioca, had been taken from its native America to Mauritius and from there it was taken in 1792 to Calcutta's botanical garden. The introduction of such edible, cultivated plants was inspired by humanitarian or utilitarian ideals, with the aim of increasing food availability.

The diffusion of edible plants is a phenomenon which has its origins in prehistory when barley and wheat, native to western Asia, spread as far as Great Britain. We know that Roman agronomists transported Mediterranean plants as far as the northern boundaries of the Empire wherever possible; the vines growing along the Rhine are an obvious example. The Arab domination of Spain also played an important part in the diffusion of such plants as sugar cane, rice and some citrus fruits. The discovery of America, however, produced an intense exchange of food resources between the Old and New Worlds, which markedly increased the resources of both. When the first Europeans landed on the other side of the Atlantic, the only plants common to Europe, Asia and Africa on the one hand and to the Americas on the other, were the coconut palm and some marrows. The process of mutual acclimatization started when Columbus brought maize back to Spain on his first voyage. When he set off again to enlarge upon his discoveries, he loaded his ships with bags of European seed grain. The purpose of his second voyage was partly to take supplies to the small colony he had left on Hispaniola after the shipwreck of the *Santa Maria*, but no one was alive when he returned.

Occasionally, the exchange of foodstuffs happened in curious ways. To the south of the Sahara, for instance, indigenous food resources consisted principally of sorghum, millet, yams, sesame and oil-producing palms; bananas are thought to have been introduced at the beginning of the Christian era, possibly by the Indonesians who invaded Madagascar. A thousand years later, the Persians introduced the mango to the coast of east Africa, but it was probably the ships of the slave trade which brought maize, manioc, sweet potatoes, runner beans and peanuts from America to Africa. The coconut palm and rice were taken to west Africa by the Portuguese trading vessels on their way back from Asia.

When the Polynesians spread from southeast Asia to the Pacific islands, they took with them the coconut palm, yam and taro and planted them in an environment where the only native plants were certain pandanuses and possibly the breadfruit tree. Even in the vast continent of Asia there is evidence of food plants being introduced—the Venetians took them to the Near East, the Portuguese to India and the Spanish to the Philippines from Mexico. By the end of the 16th century, maize had reached the Euphrates and the pineapple and papaya had reached India, to be followed a little later by sweet potatoes and peppers. Late in the 18th century, cassava reached Sri Lanka and the potato arrived in India.

The staple food of tropical America was maize, followed by cassava, sweet potatoes, peanuts and runner beans. The Spanish introduced sugar cane and bananas; rice and citrus fruits followed later. The yam was brought with the slaves from Africa. Curiously enough, in the temperate regions with a mainly white population, imported wheat has become more important than maize, and the potato, which originated on the American continent, was reintroduced into the English colonies there by the Irish settlers of the 18th century.

The pineapple

As we have seen, Christopher Columbus discovered the pineapple during his second journey to Guadeloupe. This sweet, fleshy fruit like a huge pine cone caused amazement and occasionally suspicion in Europe. Apparently Charles V refused to taste it for fear of being poisoned. The species is native to Mexico, Panama, the upper Orinoco river, Guyana and Brazil. There were no pineapples on the Hawaiian Islands when Cook discovered them, but today they produce half the world crop. The trees were introduced some time after the islands were discovered and their cultivation prospered when it was found that there was a market for the peeled, sliced and canned fruit.

The pineapples Columbus found were also cultivated. It seems that they were first grown by the indigenous tribes of South America, and their cultivation slowly spread northwards. The plants were called *nâna* by the natives, and this is the root of the generic name *Ananas*. It is still unclear whether the genus contains only one species (*Ananas sativa*) or several. Curiously enough, the pineapple's diffusion in the tropical areas of Asia (India, Sri Lanka and Indonesia) was due to human intervention, but was probably not deliberate. Pineapples last longer than many other types of fresh fruit and were therefore important as provisions on board ship at a time when the avoidance of scurvy was a major problem on long sea voyages. It may be that the first Asiatic pineapples grew from the plumes of leaves (which root easily) thrown overboard by the Portuguese sailors of the 16th century. In the 18th century, when Ehret painted the pineapple in the Chelsea gardens (called *Ananas folio vix serrato* and later *Ananas aculeatos* after Tournefort) it had become a veritable passion among the English aristocracy, who went to great lengths to cultivate the fruit in glass-houses, and their efforts were emulated elsewhere in Europe.

The banana

Of the exotic plants in the Chelsea botanic garden that Ehret drew for Christoph Jakob Trew, the banana has probably been known in the West for longer than any other. The Latin generic name, *Musa*, comes from the name of Antonio Musa, who was physician to the Emperor Augustus. Pliny mentions the banana, knowledge of which may have reached the Mediterranean through the conquests of Alexander the Great. The term used by Pliny, *pala*, can be linked to the name used in Malabar, in tropical India.

The Florentine merchant Francesco Carletti travelled the world for fifteen years on business. When he returned to Florence in 1606 he wrote a delightful travelogue entitled *Ragionamenti sopra le cose vedute né viaggi dell'Indie occidentali e d'altri paesi* (Discourses on things seen while travelling in the West Indies and other countries), which opens with a reference to the Portuguese on the

Cape Verde Islands: 'In this season [the monsoon season] they still enjoy the shade of another plant, which has very green leaves which are so large that a person can shelter beneath them. It produces fruit up to one span in length, which they call *badanas*. They are about as large as one of our cucumbers and have a smooth skin. They are peeled like our figs, but the skin is thicker and firmer. The inside is delicious to eat; it is pleasant and sweet, not too soft, a bit like a ripe melon, but drier and with no juice . . . Various dishes are prepared with them, both by the Castilians in the West Indies and by the Portuguese in the East. The latter call them *figos* and the former *platanos*, but they are in fact the same plant which is grown in countless varieties; one variety has fruits which can be eaten in one mouthful.'

The history of the banana palm is curious in several ways. It seems to have been cultivated even earlier than rice and was probably grown originally as a vegetable, in that certain Far Eastern tribes consider the hearts of the young shoots a delicacy. The genus originated in the vast tropical area which stretches from Sri Lanka to Indochina and Malaysia and includes part of Polynesia. Cultivated bananas are the product of the hybridization of two wild species; their fruits are sterile and they would not survive unless constantly replanted. It is likely that selected fruiting varieties were introduced into Africa by the Arabs. They spread throughout the continent and were also grown on the Canary Islands. From here they were introduced into the West Indies by the Spanish in 1516, while the Portuguese took them from Guyana to Brazil and from the Congo (Zaïre) to Guinea.

Jerusalem artichokes

One of the plants introduced into Europe during the 16th century was the Jerusalem artichoke. These artichokes have, in fact, nothing to do with Jerusalem; the name may be a corruption of the Italian *girasole*, meaning sunflower, since both species belong to the genus *Helianthus*. Alternatively, the name may be an indication of the route by which these vegetables reached England from South America.

The geographical discoveries of this period were mainly made by the Portuguese and the Spanish who divided the new lands between them, but Italy played an important part in the exploitation of the resulting botanical discoveries. This was partly because, as we have seen, it was in Italy that botany became established as a science, and partly because the kings of the Iberian peninsula obtained the financial backing for their expeditions from Italian bankers. In addition, agriculture was being developed to feed the expanding populations of the thriving cities of northern Italy, and the landed classes developed a passion for gardening which went hand in hand with curiosity about exotic plants. The yucca, sumach and passion flower were grown in Italian gardens in the 17th century, and the tomato (called *Amoris pomum*, meaning love-apple, because of its supposed aphrodisiac properties) had been part of the country's diet since the 16th century.

A little later peppers (genus *Capsicum*) were introduced from South America. Capsicum seeds had already been taken to the East Indies; the German botanist Leonhard Fuchs called the plants 'Calicut peppers', after the commercial port in south-west India. Aubergines had been introduced from tropical Asia and rhubarb from central Asia. Rhubarb had, in fact, been used for medicinal purposes by the Romans, who bought the dried roots, which were one of the goods transported along the Asiatic caravan routes to the trading centres of the Black Sea.

Tea, coffee, chocolate

Theobroma, which is Greek for food of the gods, was the name chosen by Linnaeus for the American cocoa plant.

The three stimulating and soothing drinks which are today part of everyday life—tea from China, coffee from Arabia and cocoa from Mexico—all became widespread in Europe during the 17th century. The descriptions in Carletti's *Ragionamenti,* written in the first years of the 17th century, of the cocoa-drinking habits among the Spanish in the New World, were therefore very much a novelty. The American plant, he wrote, 'is mainly consumed as a drink, which the Indians call *chocolat* in Nahuatl. It is obtained by mixing the fruits, which are as large as acorns, with hot water and sugar; before this, however, the fruit must be dried thoroughly, toasted over the fire and ground on special stones, just as painters do with their colours. By rubbing them with a pestle made of stone along a smooth stone, the fruit is reduced to a paste which is then diluted with water and commonly drunk by all the local people.

'Once the Spaniards and any foreigners who go there become accustomed to it, they become so addicted to it that they find it very hard not to drink it in the morning, or later in the day after lunch when it is hot, particularly while at sea. They therefore carry it in boxes, mixed with spices, or moulded into tablets which dissolve in water. This water is then put into bowls made out of large fruits, produced by local trees, which are similar to small pumpkins, except that they are round and once dry they are as tough as wood. The Spaniards drink *chocolat* in these bowls, stirring it with a piece of wood or the palms of their hands until it produces a reddish foam. They then drink it immediately, all in one go, with great relish and satisfaction. They are so strengthened, nourished and invigorated by it, that any who are accustomed to drinking it cannot keep their strength without it. Even though they may partake of something more substantial, they feel faint if they do not have their drink at their usual time.'

The custom of drinking chocolate originated among the Aztecs, who called the cocoa plant *cacaoquahuitl.* In keeping with their highly ritualized culture, they performed special ceremonies at planting time. The drink, mixed with honey and the juice of the agave, was reserved for members of the privileged classes; the others mixed cocoa with maize flour boiled in water and spiced with cayenne pepper. Cocoa seeds were also used to pay tributes; at the time of the Spanish conquest, it was recorded that Montezuma had 40,000 sacks of beans, equivalent to a million pounds in weight, in just one warehouse.

It is said that, during the colonial period, it was the nuns of Oaxaca who improved the Aztec recipe for the aristocratic drink by adding the flavours of cinnamon and vanilla to that of cocoa. However, chocolate was not popular outside Mexico, apart from in Spain. Indeed, for some time the Dutch poured into the sea the loads of cocoa beans they found on board the Spanish and Portuguese merchant ships they seized in the course of their relentless piracy.

According to the historian Fernand Braudel, Marie Thérèse, the wife of Louis XIV and daughter of Philip IV of Spain, used to drink chocolate secretly. The marriage of the Sun King had taken place in 1660. A few years earlier, the archbishop of Lyon, brother of Louis XIII's famous minister Cardinal Richelieu, had drunk it for medicinal purposes, as instructed by some Spanish nuns. The first English 'chocolate house' opened in 1657, but the drink was made very expensive by custom duties. Cocoa became fashionable in France in the 18th century after the death of Louis XIV because the Regent, Philippe, Duke of Orleans, used to drink it, mixed with milk, during his levée, in the presence of a few privileged courtiers.

However, not everyone agreed with this fashion, even at the French court. The Princess Palatine Charlotte Elisabeth of Bavaria, mother of the Duke of Orleans, wrote in 1712: 'I cannot abide tea, coffee or chocolate. I cannot even begin to understand how anybody can appreciate such things. Tea tastes to me like hay and rotten straw, coffee like soot and lupins, and as for chocolate, it is too sweet.'

Linnaeus

The help of the Dutch physician Herman Boerhaave was sought by the Tsar and the Pope, among many others. It was he who introduced the young Linnaeus to Sir Hans Sloane with this note in Latin: '*Est unice dignus te videre, et unice dignus a te videri; qui vos videbit simul, videbit hominum par cui simile vix dabit orbis . . .*' (He is the only one worthy of seeing you and of being seen by you; he who sees you together, will see two men the like of which the world is unlikely to produce again). Linnaeus visited London briefly while he was completing his medical studies in Holland.

Linnaeus had already proved himself as a botanist. In 1695 King Charles XI of Sweden, wishing to learn about and make known the natural features of his country, asked Olof Rudbeck to carry out a scientific study of Lapland's flora. Seven years after this had been finished, a fire in the library in Uppsala destroyed the herbals and manuscripts containing the results. It was not until 1732, after Sweden had recovered from Charles XII's wars, that Stockholm's Academy of Sciences was in a position to recommission Rudbeck's project, entrusting it to Linnaeus, who, at twenty-five, was already well known for his passion, if not yet for his talent, for botany. Linnaeus' fieldwork involved climbing mountains and fording rivers, dressed in Lapp costume and plagued by mosquitoes, during the nightless days of the Arctic Circle.

Carl Linné (later von Linné) was the son of a Lutheran pastor from Råshult, in Småland, a rural area in the south of Sweden. He was destined to become a clergyman himself, but he inherited his father's enthusiasm for botany rather than his religious vocation. He was encouraged by Dr Johann Rothman, a medical practitioner who taught logic and physics at the school Linnaeus attended. Rothman gave Linnaeus the works of Joseph Pitton de Tournefort, who had collected plants on the shores of the Black Sea and the Caspian Sea, in Armenia, Georgia and on Mount Ararat. Tournefort had also devised a system of classification of botanical species which was to be superseded by Linnaeus' own.

In 1735 Linnaeus went to Holland to take a doctorate in medicine at Harderwijk and to immerse himself in the atmosphere of a country then in the forefront of botanical studies. Boerhaave found him a good position as physician and scientific adviser in the house of a rich gentleman, George Clifford, who owned a botanical garden and to whom Linnaeus was to dedicate the banana, *Musa cliffortiana*. Boerhaave also offered him help and encouragement in publishing the manuscripts of theoretical works he had brought from Sweden, such as *Systema naturae*, *Fundamenta botanica*, *Genera plantarum* and his *Flora Lapponica*, giving the results of his scientific expedition to Lapland a few years earlier. The *Systema naturae* was published in Leiden in 1735 as a folio edition of only twelve pages; there have been sixteen further editions. Linnaeus spent three years in Holland and in 1738, after a trip to Paris and London, he returned to Sweden. In 1741 he was awarded the Chair of Medicine at Uppsala and spent the rest of his life there, immersed in his vast scientific studies.

Linnaeus covered all the fields of natural history, but he made two great contributions to botany. One was the introduction of binomial nomenclature which, in its simplicity and brevity, superseded the existing method of naming plants by means of a brief description (for instance, *Robinia pseudoacacia* was originally called *Arbor siliquosa Virginensis spinosa, locus nostrantibus dicta*). The second was his 'happy intuition about the importance of reproductive organs as determining characteristics in plant classification' (Sergio Tonzig). The Linnaean system of classification was called 'the sexual system' and, as A.G. Morton points out, one of the things which contributed to its popularity among the botanists of the time was the fact that the principal classes of plants were illustrated. The illustrations were executed by Georg Dionysius Ehret during his stay in Holland and were produced with Linnaeus' help as a 'tabella' which Ehret sold for two Dutch guilders a copy.

The etymology of plant names

Isidore of Seville, a Spanish prelate and saint who lived from around AD 560 to 636, wrote a vast encyclopedia known as *Etymologies*. He was considered the most learned man of his time but his work is rife with confusion. His discussion of botany covers some 230 plants. According to him, the elm, *ulmus*, is so called because it grows well in marshlands, or *uliginosis locis*. The laurel, *laurus*, is named after the Latin for praise, *laus*; the fig, *ficus*, after *fecunditas*, fertility; the juniper, *juniperus*, is linked with fire because part of its name is derived from the Greek *pur*.

And yet the true etymology of the names of certain plants is often stranger than the forced interpretations of Isidore. The carline thistle, *Carlina vulgaris*, for instance, is named after Charlemagne because of the following legend. When the Emperor contracted the plague during one of his campaigns, he prayed to God for help. God sent him one of his angels who, by shooting an arrow, showed the Emperor the herb that would heal him. A herbal dating from about 1500 depicts the plant with an arrow through its root in front of the kneeling Charlemagne, while an angel in the corner points to the plant.

Many other plants are named after eminent people, either as a compliment—for example, Linnaeus named a genus after Lord Bute—or as a form of academic recognition among botanists. The robinia was named after the French botanist Jean Robin, who introduced it into Europe in the 17th century; the fuchsia was named after the German Renaissance botanist Leonhard Fuchs; the camellia was dedicated by Linnaeus to the 17th-century Jesuit George Joseph Camellus, a missionary and pharmacist in Manila who studied the flora of the Philippines; and the name dahlia was given to these American plants by the director of Madrid's royal gardens in memory of the 18th-century Swedish botanist Dahl.

The genus *Malpighia* commemorates the Italian scientist, Marcello Malpighi; the *Gazania*, Theodoro Gaza, Theophrastus' translator; the *Thunbergia* (black-eyed Susan), Carl Peter Thunberg, one of Linnaeus' pupils who travelled round the world discovering new species. The tropical *Colum-*

naeae is named after Fabio Colonna (1567–1670) of the noble Roman family, who dedicated himself to studying herbs in order to find a remedy for the epilepsy from which he suffered—at one stage he thought he had found a remedy in valerian. The *Brunfelsia* commemorates the German botanist, Otto Brunfels, perhaps not without a hint of satire: the flowers of this plant change from violet to white as they develop, and Brunfels, a monk, changed his religion and became a Protestant at the beginning of the Reformation.

Begonias are named after the French magistrate, Michel Bégon, who brought them back from the West Indies, where he had been sent in 1681; and the poinsettia is named after Joel Roberts Poinsett, the first diplomat to be sent to Mexico by the United States government in 1825. The *Strelitzia* (*Strelitzia reginae*, or bird-of-paradise flower) was discovered in the Cape Town area by Joseph Banks, who named it after George III's wife, Charlotte, whose family name was Mecklenburg-Strelitz. *Dampiera* recalls William Dampier, the English navigator and daring buccaneer, who, as we shall see, made a contribution to botany.

Another interesting group of botanical names is derived from classical mythology. The milfoil or yarrow (*Achillea millefolium*) was used by Achilles to treat his soldiers' wounds during the Trojan War. The various *Centaurea* are so called because the centaur Chiron taught Achilles his art with herbs and used these plants to heal one of Hercules' wounds. Artemisias recall either the goddess Artemis or the Carian queen Artemisia, an expert on medicinal herbs. The latter was the loving wife of Mausolus and died of grief at his death. It has long been thought that absinthe (*Artemisia absinthium* or wormwood) causes melancholy similar to death. Surprisingly, narcissi are not in fact named after the mythical youth who was in love with his own image, but after *narké*, meaning intoxication, which is also the root of the words narcosis and narcotic.

Eupatorium cannabinum (hemp agrimony) was named in honour of Mithridates VI Eupator, King of Pontus, by his physician Krateuas; *Lysimachia nummularia* (moneywort or creeping Jennie) has rounded leaves similar to coins (Latin *nummus*); *Geranium robertianum* (herb Robert) recalls St Rupert, bishop of Worms and a phytotherapist.

According to the physician Castor Durante, *Hypericum* (St John's wort) was 'so hated by evil spirits that when it is made into poultices or burnt in houses known to be inhabited by such spirits, they immediately depart; some people therefore call it "devil chaser" and similar names'.

The cornflower (*Centaurea cyanus*) blunts the blade of a sickle; the Romans called it *baptisecula*, the Latin for sickle being *secula*, as the Frenchman Jean Ruel (Ruellius) pointed out in his *De Natura Stirpium* (1536). The greater celandine (*Chelidonium majus*) derives its name from the Greek *chelidón*, meaning swallow, but apothecaries preferred to think of its etymology as *coeli donum*, a gift from heaven, to emphasize the plant's healing properties. Similarly, the name *Centaurea* was often related to the numeral *centum* rather than to the mythical half-man, half-horse, giving rise to the German vernacular name of *hundertgüldenkraut*, or hundred-guilders-herb, which became, either through devaluation of the coin or through greater appreciation of the herb's virtues, *tausendgüldenkraut* and even, in some dialects, *milliontousendkrut*—the herb of a thousand or a million guilders.

Many of the current scientific names of plants are those given them by Linnaeus. This is indicated by the initial 'L' after the two Latin names: for example, *Nepeta cataria* L., catmint. When Linnaeus classified the cardoon and the artichoke, he called them respectively *Cynara cardunculus* and *Cynara scolymus*. In Greek, *kynara* indicates the dog-rose but the 1st-century Roman writer Columella named a kind of cardoon *Cynara*. *Carduncul*us is a diminutive of *carduus; scolymus* indicates an edible cardoon. European vernacular names for the artichoke (the French *artichaut*, German *Artischocke* and

Italian *carciofo*) derive from the Arab word *al kharshuf* or from its form without the article: *kharshuf*.

Coincidentally, the name Carl Linné, as Linnaeus was originally called, is itself derived from that of a plant. It comes from the Swedish for lime, *lind*. His Latinized name became part of biological nomenclature with *Linnaea borealis*, a genus of the Caprifoliaceae. Linnaeus dedicated the *Lobelia*, a genus of the Campanulaceae, to de Lobel, whose name actually means poplar, and, as a chivalrous gesture, he called a species of iris *Moraea*, after his own wife, Sara Moraeus.

Travelling botanists

In the 15th century, an unidentified rich German wrote: 'But when, in the process of the work, I turned to the drawing and depicting of herbs, I marked that there are many precious herbs which do not grow here in these German lands, so that I could not draw them with their true colours and form, except from hearsay. Therefore I left unfinished the work which I had begun, and laid aside my pen, until such time as I had received grace and dispensation to visit the Holy Sepulchre, and also Mount Sinai, where the body of the Blessed Virgin, Saint Catherine, rests in peace. Then, in order that the noble work I had begun and left incomplete should not come to naught, and also that my journey should benefit not my soul alone, but the whole world, I took with me a painter ready of wit, and cunning and subtle of hand. And so we journeyed from Germany through Italy, Istria, and then by way of Slavonia or the Windish land, Croatia, Albania, Dalmatia, Greece, Corfu, Morea, Candia, Rhodes and Cyprus to the Promised Land and the Holy City, Jerusalem, and thence through Arabia Minor to Mount Sinai, from Mount Sinai towards the Red Sea in the direction of Cairo, Babylonia, and also Alexandria in Egypt, whence I returned to Candia. In wandering through these kingdoms and lands, I diligently sought after the herbs there, and had them depicted and drawn, with their true colour and form' (A. Arber, *op. cit.*). Thus pious intentions and a taste for discovery combined to give rise to one of the first botanical journeys. It resulted in the herbal *Gart der Gesundheit* (Garden of Health, known in Latin as *Hortus Sanitatis*) printed in Mainz in 1495 by Peter Schöffer, Gutenberg's partner and successor.

Other itinerant botanists took an interest in the plants of the Levant. One was Prospero Alpino, who published his *De Plantis Aegypti Liber* in 1592. At the end of the 16th century, after graduating, Alpino had become general practitioner to the little community of Campo San Pietro, a castle near Padua, within the Venetian political boundaries. He was not there long, for he was appointed physician to the Republic's consul in Egypt. During the three years Alpino spent there, he spared neither effort nor diligence in carrying out research and acquiring a full knowledge of the properties of the rarest plants that grow there. When he returned, Alpino introduced Egyptian species into the Paduan botanic garden (among them Goethe's palm). It seems that he was also the first to mention coffee, which he had seen in Cairo.

Another book by Alpino on exotic plants was posthumously published by his son. In it he mentions a botanical expedition undertaken by Giuseppe Benincasa (a Fleming whose original name was Goodenhuyse), director of the Pisan garden, under the aegis of Grand Duke Ferdinand I. From 1591 to 1593 he explored Crete, then still under Venetian rule. 'Having toured the whole island, he brought back to Italy a great number of plants with which he enriched and marvellously adorned the Pisan garden.' Benincasa was a correspondent of the Flemish botanist Clusius, to whom he wrote that he had seen a *Cyclaminum albo flore* (white-flowering cyclamen) blossoming in November.

Precious cargoes

The Florentine merchant mentioned above, Francesco Carletti, addressed Grand Duke Ferdinand in the second volume of his *Ragionamenti* (his account of his 'journey from Japan to China and matters pertaining to that kingdom'—although all that he saw of China was Macao): 'They also have another fruit which they call lychee, as excellent from the point of view of both taste and health as are many others in those countries; one can, and indeed one does, eat as many of them as one likes, since one never tires of them, nor do they make one ill. They are as big as plums, with a hard, rough skin like that of the fruit of the strawberry tree, and of a very similar red and green colour. They can be peeled very easily. The pulp is a little like an unripe grape, not too sweet, watery, refreshing and extremely pleasant indeed. They squeeze lychees to make wine, which is also delicious. In the middle of the fruit is a large stone, like a big olive. It is a dark tan colour and the skin and kernel are similar to an acorn. I brought back a large quantity of them, hoping that, once sown, they might produce even here such handsome fruits. They grow in bunches on the branches and are a curious and wonderful sight.'

It was erroneously thought that the lychee, which so delighted the Florentine, was the fruit of one of fifty or so species of the genus *Annona* (or *Anona*), most of which are native to tropical America and had been imported into Asia where the climate permitted. A specimen of this genus was drawn by Ehret for Christoph Jakob Trew. Species of *Annona*, which include the custard apple and the cherimoya, are now cultivated in tropical areas both for their fruit and for their ornamental value.

Carletti's hope of introducing the lychee into Tuscan gardens came to nothing. On his way home he spent some time in Gôa and then boarded the Portuguese carrack *Santiago*, a ship with its own little niche in history. The *Santiago* was seized in 1602 off St Helena by two Dutch ships engaged in the war of independence against Spain and Portugal, then under the Spanish crown. The war began in 1581 and lasted for 80 years. The *Santiago* was carrying a shipment of oriental porcelain, as was another Portuguese carrack, seized the following year off the Indian coast. In Amsterdam the sale of *Kraak-porcelein* (for a long time the Dutch called the blue and white Ming ware 'carrack porcelain' as a result of these incidents) was not only extremely successful but also marked the beginning of the fashion for oriental ceramics which spread over the whole of Europe.

When the *Santiago* was captured, all of Carletti's merchandise was also seized. The resulting lawsuit was heard before the 'most eminent and generous prince and lord, Maurice, Prince of Orange, Count of Nassau, admiral of the sea'. The defendant was the Chamber of Zeeland (the attacking ships came from this province) of the Dutch East India Company. Carletti was found guilty and all his goods 'impounded in the interests of the common cause and of those who have a right to them'. Carletti appealed against 'such an unjust and iniquitous sentence pronounced by judges who were themselves the defendants, one of whom even said that instead of taking me to

Zeeland they should have cast me into the sea', but this aroused little sympathy and his appeals came to nothing.

Plants and sailors

The *Ficus benghalensis* grows aerial roots from its branches to the ground. These thicken to form new trunks, like columns, so the tree grows like a huge pergola under which, in India, merchants gather to carry out their business. It is known as the banyan tree, *banyan* meaning merchant. We know from the works of Flavius Arrianus, a 2nd-century Greek historian from Bithynia who was a contemporary of the Emperor Hadrian, that Nearchus had accurately described the banyan tree in the 4th century BC, as well as several species of mangrove and the prickly euphorbia of Baluchistan. Nearchus was a Macedonian officer under Alexander the Great. After the Indian campaign, while Alexander marched his soldiers across the Iranian deserts, Nearchus led the fleet, which had been built in the estuary of the Indus, over the Indian Ocean and up the Arabian Gulf to the mouth of the Euphrates. He was the first sailor to be of service to botany.

In the 16th century, the German botanist Leonhard Fuchs wrote: 'But there is no reason why I should dilate at greater length upon the pleasantness and delight of acquiring knowledge of plants, since there is no one who does not know that there is nothing in this life pleasanter and more delightful than to wander over woods, mountains, plains garlanded and adorned with flowerlets and plants of various sorts . . . But it increases that pleasure and delight not a little, if there be added an acquaintance with the virtues and powers of these same plants' (A. Arber, *op. cit.*). This spirit survived over the centuries. We have already noted how Clusius corresponded with Busbecq in Constantinople and with Benincasa in Crete. Clusius had made a botanical journey to Spain and Portugal with two friends; in London he was able to acquire a collection of American plants thanks to the kindness of Sir Francis Drake.

Later on, the religious orders came to be of importance. Van Rheede tot Drakenstein was appointed Governor of Batavia, now Jakarta, by the Dutch East India Company, whose drastic methods have already been observed in Carletti's misadventure. Drakenstein has earned his place in the history of botany by promoting the study of exotic flora with the help of a Carmelite friar, the Neapolitan Fr Matteo di San Giuseppe, and of the pastor Casacarius. The result was the *Hortus indicus malabaricus*, published in parts in Amsterdam between 1678 and 1703. One of the commissions undertaken by Georg Dionysius Ehret, before he met Trew, was to colour the copper engravings for this gigantic work in accordance with the descriptions in the text, for the Regensburg merchant Loeschenkohl, who employed him as a gardener.

The Latin name of the Japanese iris is *Iris kaempferi*, after the German physician and botanist Englbrecht Kaempfer. Between 1683 and 1693 he travelled from Germany to Sweden and through Russia, Georgia and Persia to the Far East. Having accompanied a Swedish ambassador to Isfahan, the capital of Persia, he spent some time there and then continued along the coasts of Arabia and India as far as Java. He then sailed on to Japan. In those days, the only Europeans to be allowed to set foot in the remotest country in the world were the Dutch, and even they were confined to the tiny island of Deshima in the bay of Nagasaki. Kaempfer, like several other Germans, was in the service of the Dutch East India Company.

Many of his botanical observations were undoubtedly carried out in secret, because the Japanese at the time of the Tokugawa shogunate were acutely xenophobic. Nonetheless, his *Amoenitatum exoticarum* (On exotic delights), published in Westphalia in 1712, described things never seen by European eyes and was immensely successful. As we have seen, Sir Hans Sloane bought his manuscripts.

A network of travelling botanists grew up around Sir Hans Sloane and John Ray, the most influential English botanist of the second half of the 17th century who has sometimes been called the English Pliny. Sloane passed on the results of the botanizing he had carried out whilst a student in France to Ray, and the latter was also sent plants from the Philippines by George Joseph Kamel. A Jesuit missionary, Kamel discovered strychnine, which is extracted from the seeds of the nux vomica, an East Indian tree. The botanist William Sherard was a pupil of Ray's who corresponded with Sloane and collected plants in the Levant when he was sent as the English Consul General to Smyrna in the Turkish Empire. Another person to contribute specimens to Sloane's herbarium and to the botanical garden in Chelsea was William Dampier.

William Dampier (1652–1715), an orphan sent to sea to earn a living, led an extremely adventurous life, and became a self-taught naturalist. By the age of twenty-five he had already sailed to Newfoundland and to Java, fought in a war against the Dutch, been to the Caribbean, worked with the woodcutters in Campeche and sampled the life of a buccaneer. After a short stay in England, he went back to Jamaica. He joined a band of buccaneers engaged upon sacking the Pacific coast of South America as far as the Juan Fernández Islands.

In 1683 he enrolled in Virginia with a privateer (a seaman entitled by royal consent to attack enemy, that is, Spanish ships). The privateer was called Captain Cook but he had no connection with his famous namesake. They sailed along the coast of Guiana, rounded Cape Horn, sacked and looted in the waters of Chile and Peru, and reached the Galapagos Islands and Mexico. Here Cook died and a Captain Davis took command. After joining other English and French pirate ships, Davis concentrated on the classic target of the Spanish towns on the Pacific coast. Dampier then joined the ship of Captain Swan, who planned to cross the Pacific and return to Europe via the East Indies, with the purpose, naturally, of lining the coffers of the British monarch.

Subsequent events become increasingly like a story of derring-do: there was a threat of mutiny at Guam, debauchery in the Philippines, the marooning of Swan and other men in Mindanao, raids along the China coast and to the Spice Islands and New Holland (Australia). Finally, in 1688, Dampier himself was abandoned on the Nicobar Islands with two Englishmen, a Portuguese and a few Malayans. According to Dampier, this was at his own request, as he wished to enter the trade in ambergris. The group of men reached Sumatra in a canoe, often risking an ignominious end through hunger or drowning.

After a few more journeys and a half-hearted term as a gunner in one of the East India Company's forts, Dampier boarded one of the Company's vessels and came back to England in 1691. He had been away for eight consecutive years, but on his return became a friend of Sloane and wrote several accounts of his many adventures. He died in 1715, but during the intervening years this 'great filibuster', the 'king of the sea' who was idolized in sailors' taverns, became a respectable gentleman.

In 1699 the Admiralty made him captain of the *Roebuck* and sent him on an exploratory expedition to Australia; this was of some geographical importance but ended with the shipwreck of the *Roebuck* on Ascension Island in 1701. Dampier then commanded a privateering expedition to the South Seas and from 1708 to 1711 he was pilot under Captain Woodes Rogers on a voyage around the world, in the course of which the Scottish sailor Alexander Selkirk was rescued from one of the tiny deserted Juan Fernández Islands. The prototype of Robinson Crusoe, he had been left there by Dampier after a quarrel five years previously, with a knife, a gun and a keg of gunpowder.

A meeting on the ocean

In 1749 another botanical explorer, Michel Adanson, arrived in Senegal, where the French had a commercial outpost—a port of call for loading provisions, with a small garrison, surrounded by tropical forest. Adanson had been found a modest post working in the warehouses of the Compagnie du Sénégal by the de Jussieu brothers. A botanist from Aix-en-Provence, he was only twenty years old. As the French naturalist Georges Cuvier later said in his commendation of him at the Institut de France: 'Senegal was, of all the European settlements, the most difficult to penetrate, the hottest, the most unhealthy and dangerous in all other respects, and therefore the least known to naturalists.' Adanson spent five years there, collected vast quantities of plants and made known to Europe the giant of the plant world, the baobab, a botanical relative of the humble mallow.

At about the same time (1748–49), the Swedish botanist Peter Kalm—one of Linnaeus' pupils—visited New England and Canada, and wrote a large work on the botany, ethnography and economy of the regions. Exploration of South America was undertaken by Charles Marie de La Condamine, who led a French scientific expedition to Peru and the Amazon from 1735–45. One of the members of the expedition was Joseph de Jussieu, who remained in Latin America until 1771, in order to study the flora. This had not been the main aim of the expedition, which was more concerned with physics.

Over seventy years earlier, the physicist Richer had visited Guyana and noticed that the pendulum he had brought with him did not mark the seconds accurately, in that it was slower than it had been in Paris. This led him to deduce that the earth could not be a perfect sphere, but must be flattened at the poles and thicker around the equator. In order to have scientific proof of Richer's theory, the French Academy of Science decided that the meridional arc should be measured at different points of the globe. La Condamine was to measure it in the Andes, near Quito, which is almost exactly on the equator. This was a splendid opportunity to carry out extensive research in the fields of geography and the natural sciences. The most notable part of the expedition was probably the journey down the Amazon, as it represented the first scientific exploration of this huge waterway and of the lands, flora, fauna and inhabitants of its banks, although the very first explorer to travel from the Andes to the Atlantic was the Spanish soldier Francisco de Orellana, in 1541.

A subsequent British expedition is interesting for two main botanical achievements. The *Dolphin*, under Samuel Wallis, the *Swallow*, under Captain Carteret, and their supply ship the *Prince Frederick*, left Plymouth in 1766. The *Prince Frederick* took several thousand young plants, mainly arboreal varieties, to the Falkland Islands, which were quite barren. When Wallis reached Tahiti (he and his crew were the first Europeans to land on the island) he gave the islanders chickens and other domestic animals as well as young orange and lemon trees, which flourished.

The *Swallow* had become separated from the other vessels and went her own way. As was customary, Carteret left a bottle with a message on Ascension Island, to mark his halt there and his course. On the way back, the *Swallow* was hailed from the deck of a French frigate which had found the message on the island. It was commanded by Louis Antoine de Bougainville, who was completing his journey round the world.

An idea of the working conditions faced by botanists who, for the love of science or of adventure, embarked on voyages lasting several years aboard the handsome yet uncomfortable sailing ships of the 18th century, is given by an anecdote in Bougainville's *Journey around the world*. Louis Antoine de Bougainville left Brest on the *Boudeuse* in 1776. The ship's botanist was Philibert Commerson, a young man who had come to the notice of Linnaeus and whom Voltaire wished to employ as his secretary. When they reached the coast of South America, Commerson dedicated the bougainvillaea—the beautiful climber with a lovely scent which came to be cultivated all over the world—to his captain.

Commerson had an indefatigable assistant named Baré, who was an experienced botanist himself. Bougainville describes how Baré accompanied Commerson on all his botanizing trips, in deep snow and on the ice-covered mountains of the Straits of Magellan. On those exhausting marches he always carried the provisions, weapons and botanical notebooks, and his strength and courage earned him the nickname of 'beast of burden'. It transpired that Baré was, in fact, a woman who had boarded the ship disguised as a man. Her secret was discovered by Bougainville, who believed the moving story the girl, in tears, told him: she was an orphan, had been reduced to penury after losing a lawsuit, and was consumed with curiosity about the world. Reality was somewhat different. Jeanne Baré had been employed for years by Commerson as a governess; she looked after his house and herbaria, and continued to serve him when he settled on the Ile de France (Mauritius), where he died four years later. She then married a blacksmith.

Jeanne Baré's story is also related in Diderot's *Supplément au voyage de Bougainville*, which takes the form of a dialogue between the two anonymous characters, A and B.

'B: She was born in Burgundy and was called Baré. Neither plain nor pretty, at twenty-six she had never left her village. When she first decided to travel, it was to go round the world. She always showed good sense and courage.

A: These fragile human contraptions sometimes contain souls full of fortitude.'

The dialogue refers thus to Bougainville:

'A: I really cannot understand that man. He spent his youth studying mathematics, which presupposed a sedentary life; and then suddenly he gave up a life of meditation and concentration for the active, arduous, itinerant and dissipated life of the traveller.

B: Not at all. When you consider that a ship is but a floating house, and that a navigator crosses vast expanses, all the while closed in a very restricted space, you will realize that he goes round the world on a wooden plank as you or I go round the universe while standing still on our own floor.'

Botany Bay

A quotation from the log-book of the *Endeavour*, the ship on which Captain Cook made his first journey, dated Sunday, 6 May 1770, off the eastern coast of Australia, near the place where Sydney was later to be built: 'In the evening the yawl returned from fishing, having caught two Sting-rays weighing near 600 pounds. The great quantity of New Plants Mr Banks and Dr Solander collected in this place occasioned my giving it the name of Botany Bay. It is situated in the latitude of 34°0′ S and longitude 208°37′ W; it is capacious, safe and commodious . . . We set anchor by the southern shore, about a mile inside the mouth, to be able to set sail with the wind in the south and find fresh water . . . Although there is plenty of wood, the variety is not great: the largest trees are as large or larger than our oaks in England and grow a good deal like them, and yield a reddish gum; the wood itself is heavy, hard and black like Lignum Vitae; another sort that grows tall and strait something like Pines, the wood of this is hard and ponderous and something of the nature of American live oaks; these two are all the timber trees I met with. There are a few sorts of Shrubs and several Palm trees and mangroves about the head of the harbour . . .'

Cook was accompanied on his journeys by at least three of Linnaeus' disciples. Banks and Solander sailed on the *Endeavour*. Joseph Banks was then twenty-five, had a private income of £6,000 a year, was a Fellow of the Royal Society and had already been on a botanical expedition to Labrador and Newfoundland. He was the scientific director on Cook's voyage of discovery, which was the first expedition to include an organized group of scientific researchers. Nobody, it was said at the time, had ever set to sea better equipped to study natural history. Daniel Charles Solander left his native Sweden and settled in England on Linnaeus' advice. He was thirty-five when he embarked on the expedition and was probably more experienced than the 'young gentleman' he accompanied. Together they carried out their task splendidly.

Shortly after his return to England, Cook set sail again in the *Resolution* and the *Adventure*. Banks and Solander stayed behind in London in order to publish the results of their scientific research, but Banks nominated two other naturalists to travel with Cook: Johann Rheinhold Forster and his son Johann Georg. The father was not a sociable man, but he always had a ready answer. When Frederick II of Prussia asked him how many sovereigns he had met in his life, he answered: 'Seven, sire, four wild and three acclimatized ones.'

When the two ships stopped at Cape Town, the group of scientists was reinforced by another botanist who, like Banks and Solander, was a follower of Linnaeus. The two Forsters prevailed upon the captain to take the Swedish doctor Sparrman aboard. He had gone to southern Africa because of the descriptions he had heard of the botanical garden in Cape Town and had supported himself by becoming a tutor to the wealthy colonialists who had settled there, while pursuing his passion for botany.

Banks later became President of the Royal Society. He died in 1820, his last years troubled by gout. We have already mentioned that he was made director of Kew Gardens. In order to enrich them with a constant supply of new species, whenever the opportunity arose he sent botanists and gardeners on expeditions around the world. One of the most unfortunate was a young gardener at Kew, David Nelson, who set sail with Cook in 1776, on what was to be the captain's last voyage—he was killed on a Hawaiian beach in 1779. Nelson travelled as assistant to the surgeon and botanist William Anderson. He was one of the first naturalists to study the flora of the Aleutian Islands and the *Ranunculus nelsonii* is named after him. Eleven years later, Captain Bligh invited him to join the *Bounty*. David Nelson did not join the mutineers, and was among the eighteen men abandoned in an open boat with a couple of knives and few provisions. He survived the long and troubled journey to Timor, but died the day after the survivors landed safely.

By now Trew and Ehret had also died; almost all the wonders of nature had been catalogued and the days of wonderment were about to end, but their books remained as a testimony to the discoveries of the age of exploration.

The kitchen garden and the wood

. . . a good-looking old man, who was taking the air at his door, under an alcove formed of the boughs of orange-trees. Pangloss, who was as inquisitive as he was disputative, asked him what was the name of the mufti who was lately strangled. "I cannot tell," answered the good old man; "I never knew the name of any mufti, or vizier breathing . . . I never inquire what is doing at Constantinople; I am contented with sending thither the produce of my garden, which I cultivate with my own hands." After saying these words, he invited the strangers to come into his house. His two daughters and two sons presented them with divers sorts of sherbet of their own making; besides caymac, heightened with the peels of candied citrons, oranges, lemons, pineapples, pistachio nuts, and Mocha coffee unadulterated with the bad coffee of Batavia or the American islands.

Voltaire, *Candide*, I,XXXX,
translated by William F. Fleming, 1927

Strawberry

Fragaria vesca, Rosaceae

Linnaeus derived the generic name from the Latin *fraga*, meaning fragrant berry. Strawberries are mentioned by Virgil and Pliny, and at the beginning of the 17th century Mattioli wrote that *'fraghe'* are not only very pleasant in summer but are also thirst-quenching and very useful for 'choleric' stomachs. The plant has many traditional medicinal applications: it was used to strengthen the gums, heal wounds and stop haemorrhages.

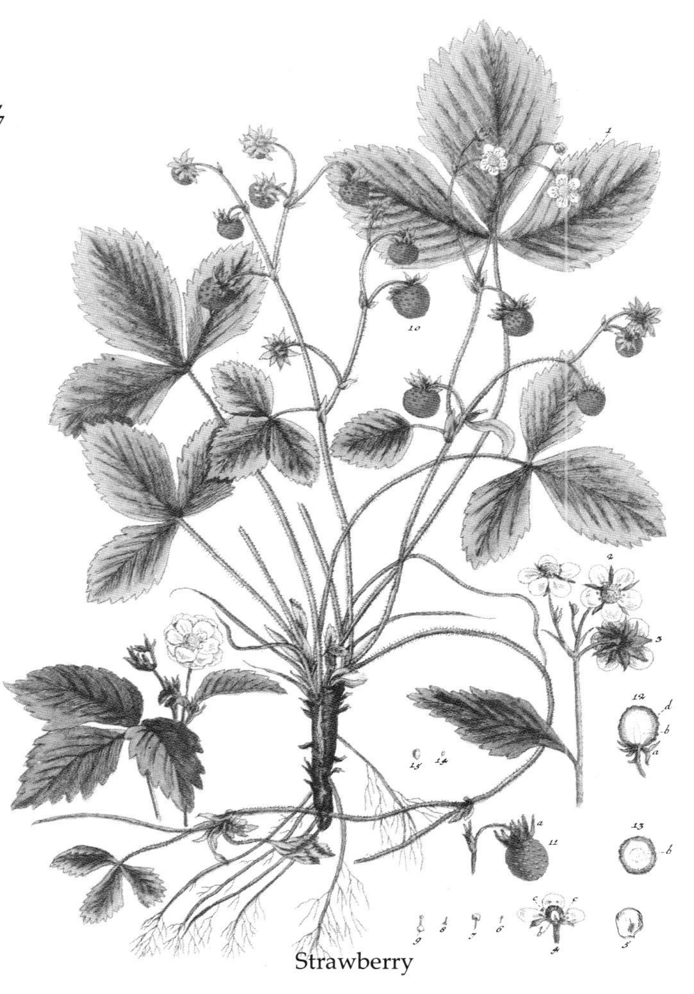

Strawberry

Fennel

Foeniculum vulgare, Umbelliferae

A literal translation of the Latin generic name would be 'small hay', probably with reference to the plant's finely cut leaves which resemble grass. Fennel is commonly found growing in dry, sunny places, particularly around the Mediterranean. It was known in classical times and was much used as a flavouring (the seeds in particular) and as a medicinal plant with sedative properties.

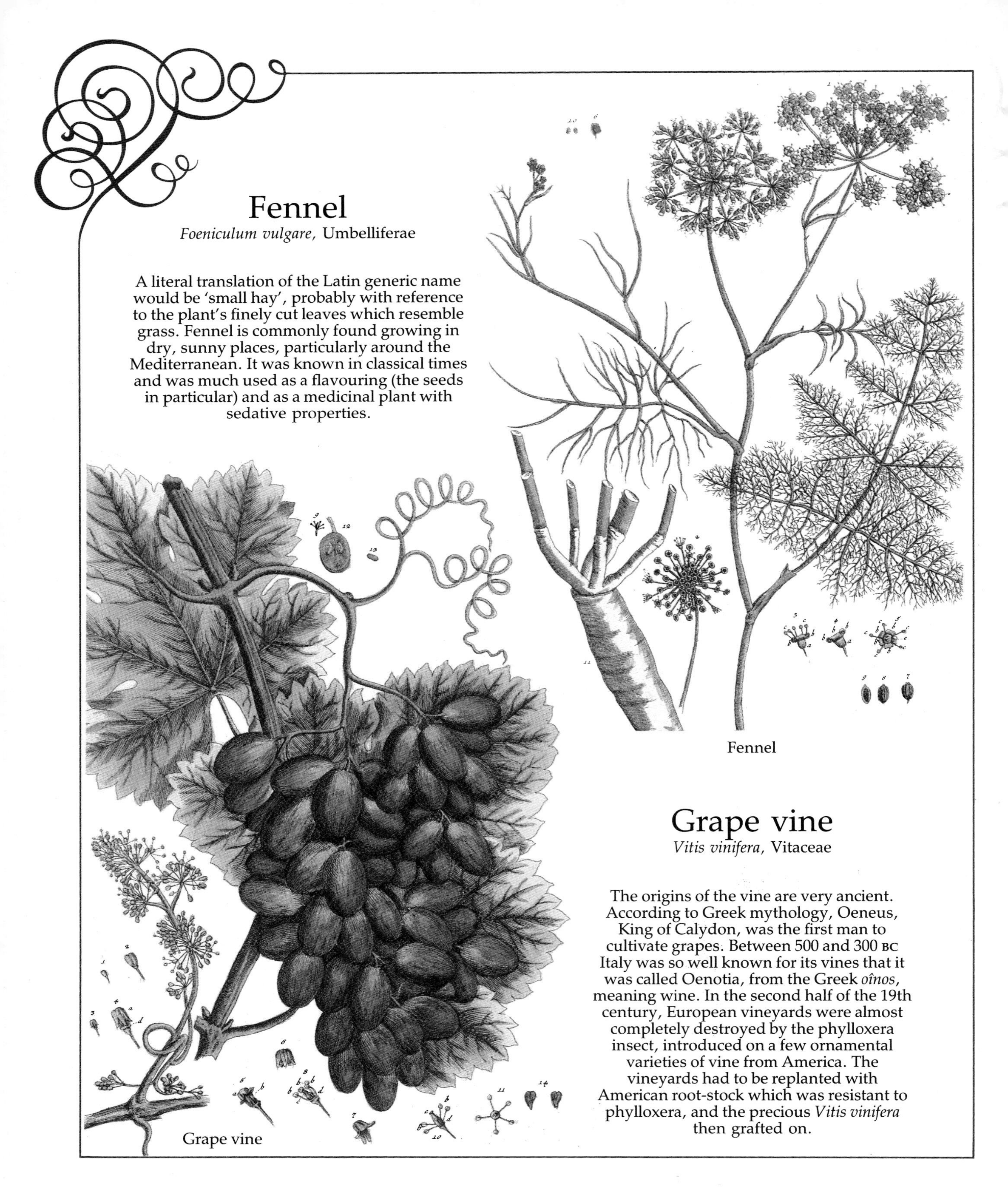

Fennel

Grape vine

Grape vine

Vitis vinifera, Vitaceae

The origins of the vine are very ancient. According to Greek mythology, Oeneus, King of Calydon, was the first man to cultivate grapes. Between 500 and 300 BC Italy was so well known for its vines that it was called Oenotia, from the Greek *oînos*, meaning wine. In the second half of the 19th century, European vineyards were almost completely destroyed by the phylloxera insect, introduced on a few ornamental varieties of vine from America. The vineyards had to be replanted with American root-stock which was resistant to phylloxera, and the precious *Vitis vinifera* then grafted on.

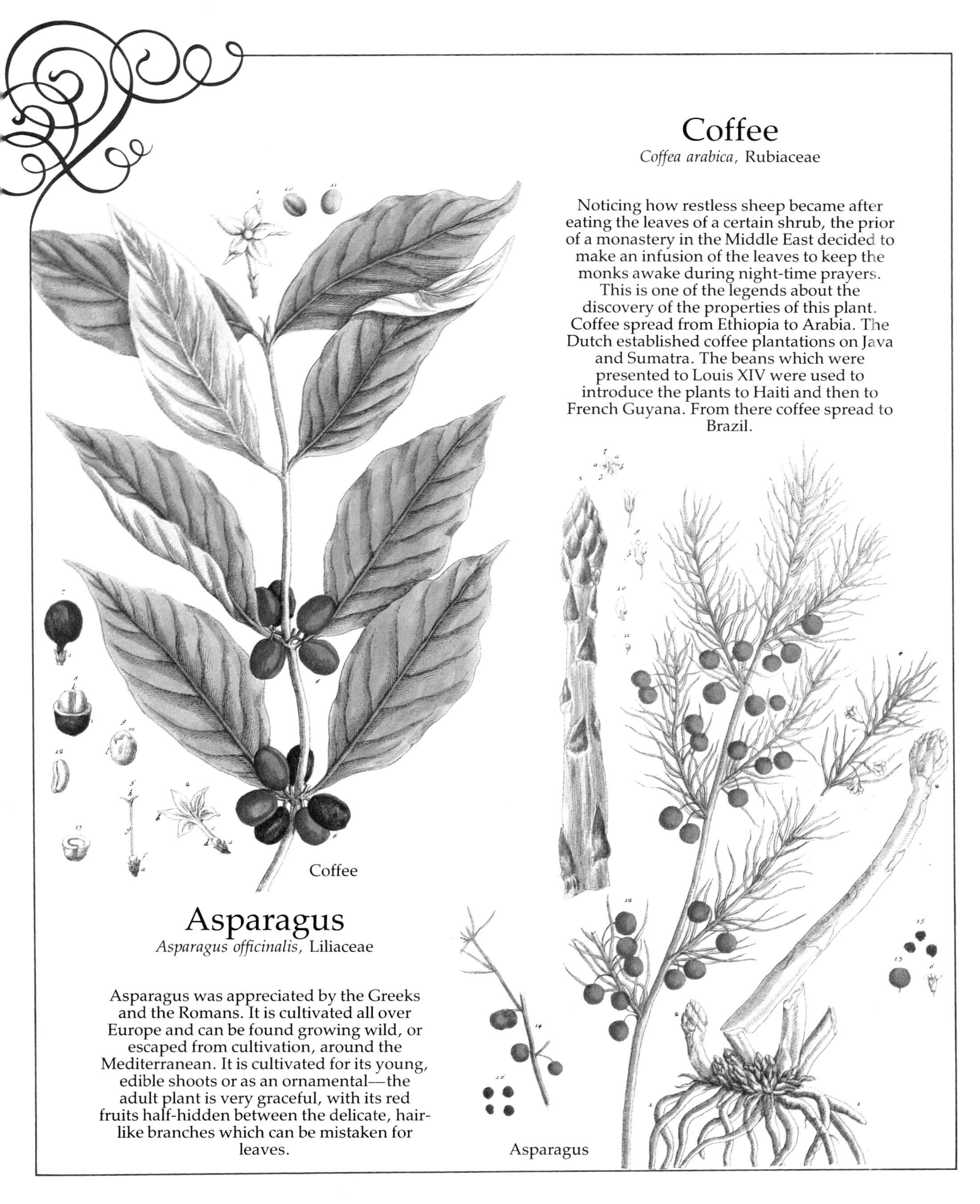

Coffee

Coffea arabica, Rubiaceae

Noticing how restless sheep became after eating the leaves of a certain shrub, the prior of a monastery in the Middle East decided to make an infusion of the leaves to keep the monks awake during night-time prayers. This is one of the legends about the discovery of the properties of this plant. Coffee spread from Ethiopia to Arabia. The Dutch established coffee plantations on Java and Sumatra. The beans which were presented to Louis XIV were used to introduce the plants to Haiti and then to French Guyana. From there coffee spread to Brazil.

Coffee

Asparagus

Asparagus officinalis, Liliaceae

Asparagus was appreciated by the Greeks and the Romans. It is cultivated all over Europe and can be found growing wild, or escaped from cultivation, around the Mediterranean. It is cultivated for its young, edible shoots or as an ornamental—the adult plant is very graceful, with its red fruits half-hidden between the delicate, hair-like branches which can be mistaken for leaves.

Asparagus

Tomato

Lycopersicon esculentum, Solanaceae

The tomato was introduced into Europe by the Spanish, either from Mexico or from Peru. For over a century, it was only used in cooking to a very limited extent, although it was known that it could be eaten like an aubergine. Its cultivation did not become widespread until the beginning of the 18th century. The English vernacular name is derived from the Mexican *tomatl*. It used to be called the love-apple because of its supposed aphrodisiac properties.

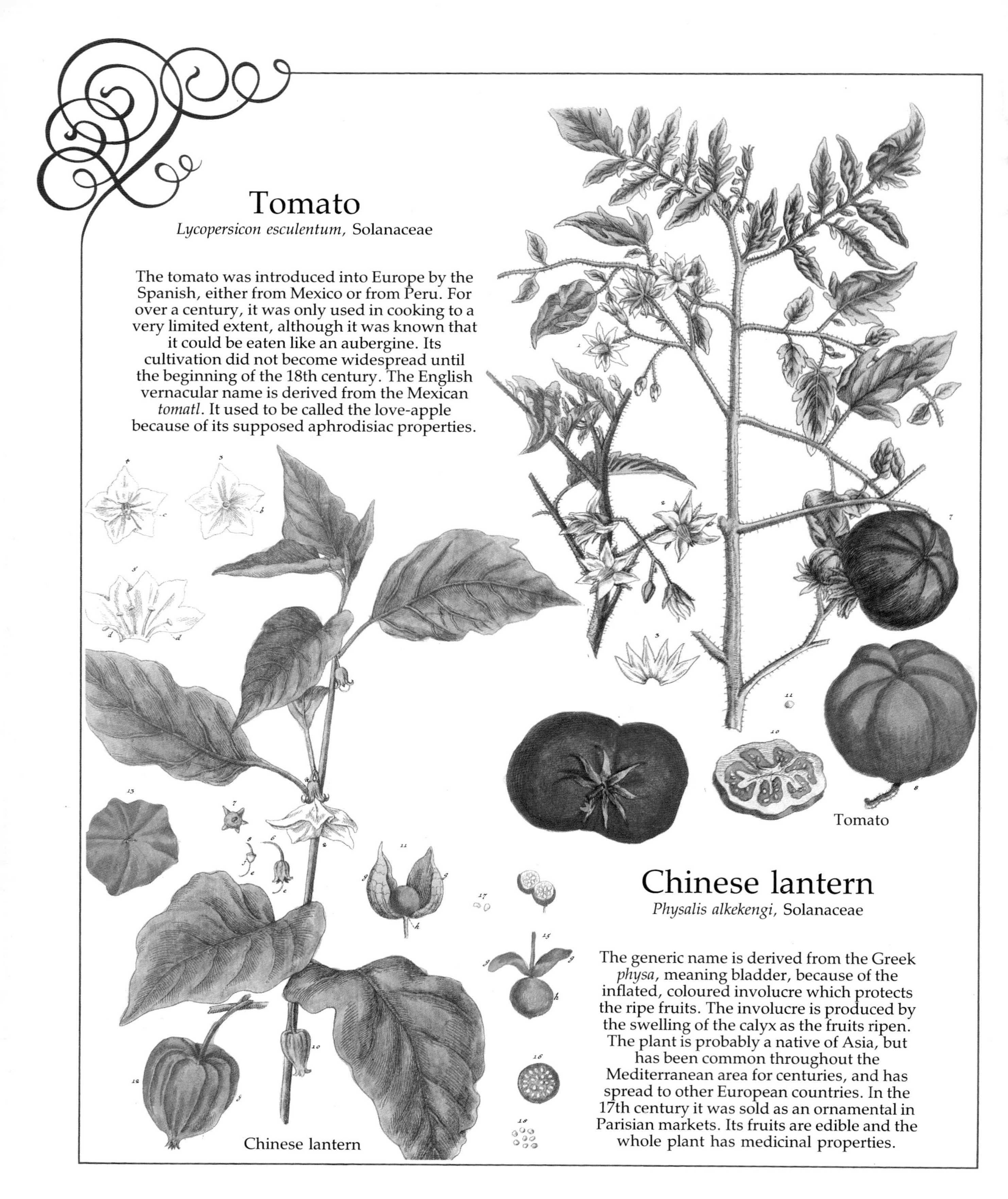

Chinese lantern

Physalis alkekengi, Solanaceae

The generic name is derived from the Greek *physa*, meaning bladder, because of the inflated, coloured involucre which protects the ripe fruits. The involucre is produced by the swelling of the calyx as the fruits ripen. The plant is probably a native of Asia, but has been common throughout the Mediterranean area for centuries, and has spread to other European countries. In the 17th century it was sold as an ornamental in Parisian markets. Its fruits are edible and the whole plant has medicinal properties.

Orange

Citrus sinensis, Rutaceae

The fruits of citrus plants are also called hesperidia after one of Hercules' labours: in order to take the golden apples of the Hesperides to Eurystheus, Hercules had to kill the dragon with a hundred heads which guarded the garden where they grew. *Citrus sinensis* is the sweet orange and *Citrus aurantium* the Seville orange.

Orange

European larch
Larix decidua, Pinaceae

The European larch is the only conifer to shed its leaves with the first winter frosts. In the autumn, the golden colour of the larch stands out against the dark green of the spruces and other conifers that are often found growing with larches. They yield a good supply of resin which is employed in the production of turpentine.

European larch

Norway spruce

Norway spruce
Picea excelsa, Pinaceae

The Romans called the spruce *picea*, and the name came to be applied to the whole genus. The Norway spruce is a majestic tree which grows wild from the Apennines to Scandinavia. In central and northern Europe and in Asia it forms vast forests and is important for timber production. Its trunk is scored with channels in which resin flows. The wood has been used to make the best musical instruments, such as the famous Stradivarius violins.

Stone pine

Pinus pinea, Pinaceae

A Mediterranean pine producing seeds with edible kernels, the god Pan wore its branches as a garland; his lover Pitus may have given rise to the name *pinus*, but the etymology is uncertain. The stone pine was sacred to the nature goddess Cybele.

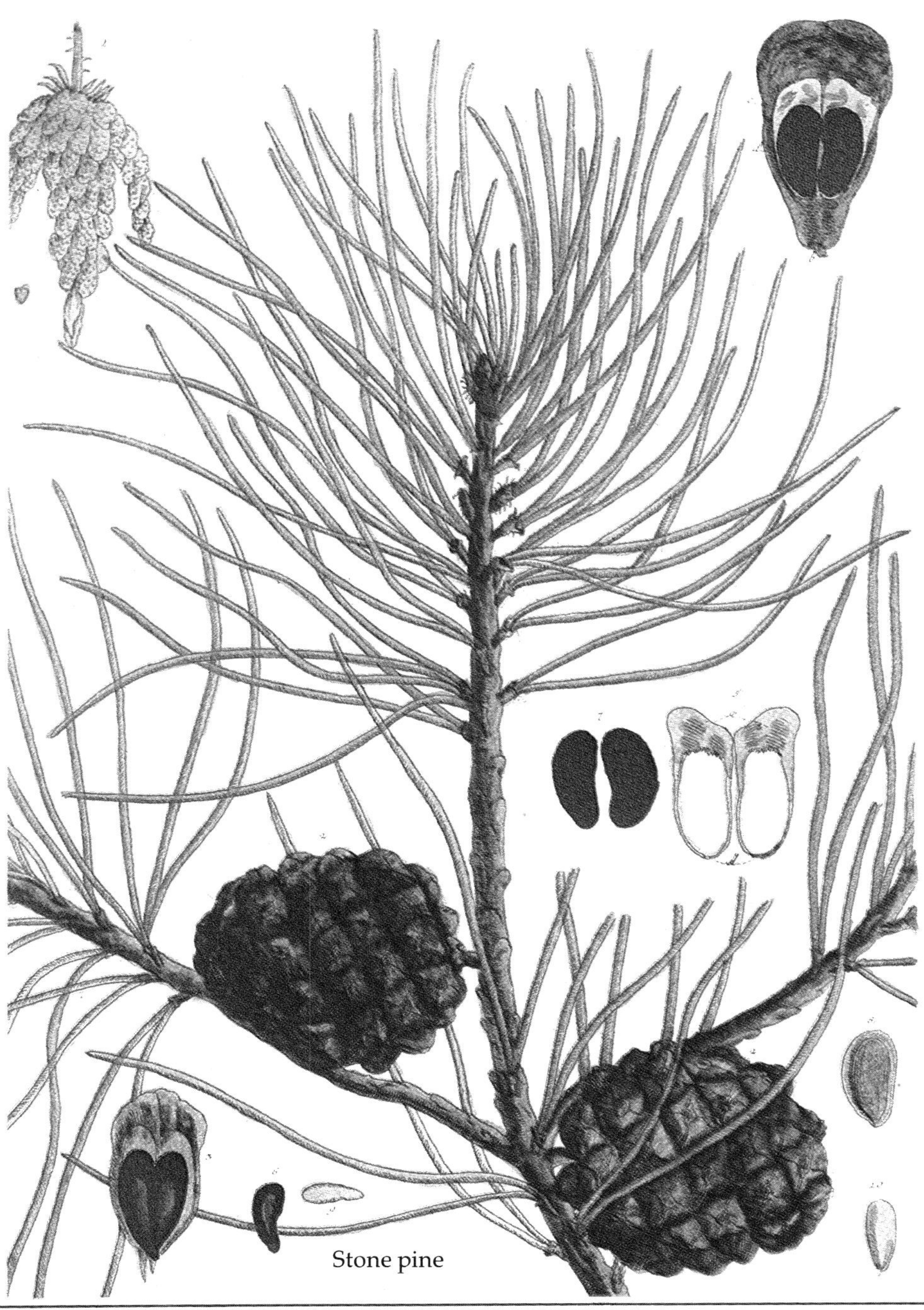

Stone pine

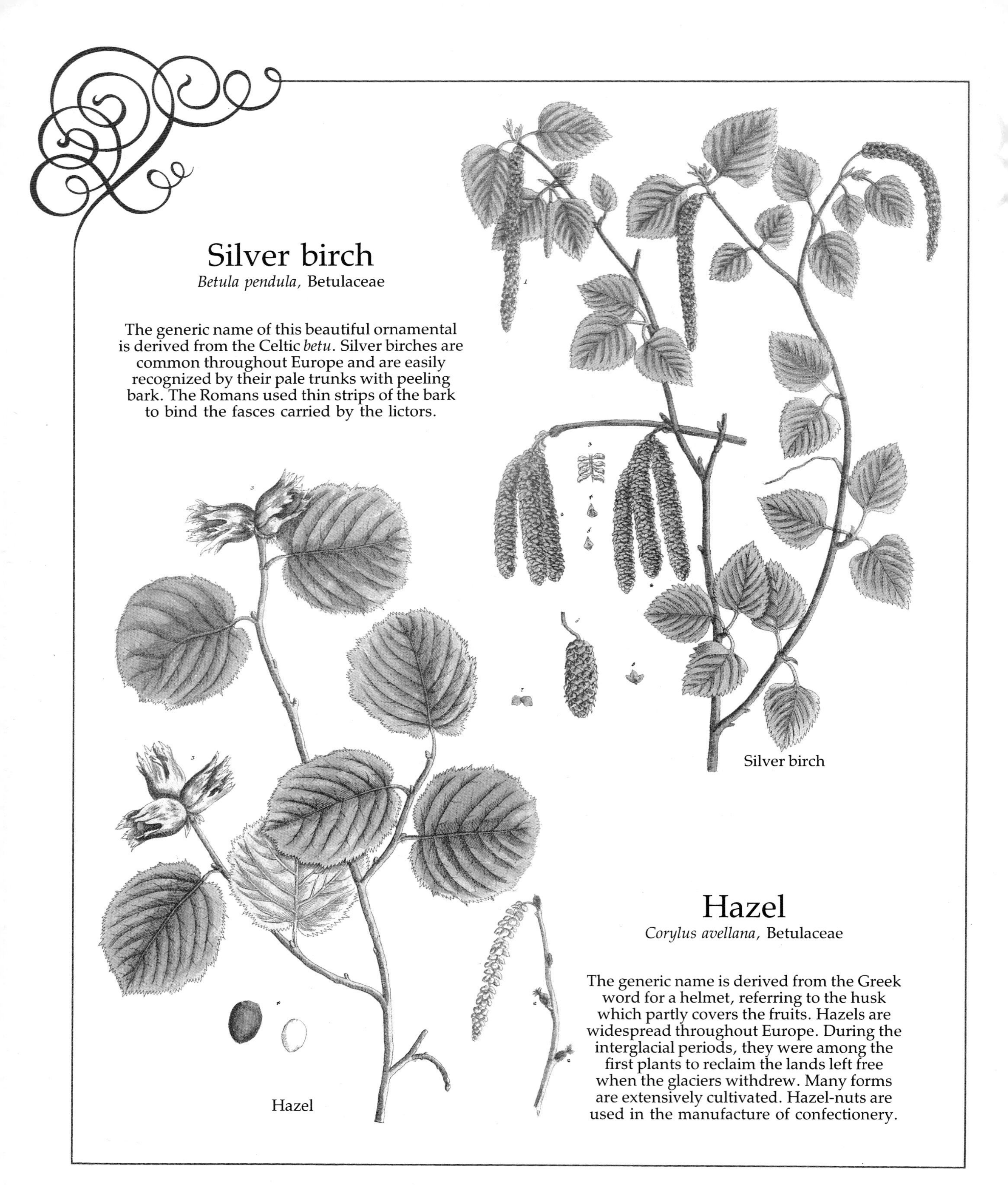

Silver birch

Betula pendula, Betulaceae

The generic name of this beautiful ornamental is derived from the Celtic *betu*. Silver birches are common throughout Europe and are easily recognized by their pale trunks with peeling bark. The Romans used thin strips of the bark to bind the fasces carried by the lictors.

Silver birch

Hazel

Hazel

Corylus avellana, Betulaceae

The generic name is derived from the Greek word for a helmet, referring to the husk which partly covers the fruits. Hazels are widespread throughout Europe. During the interglacial periods, they were among the first plants to reclaim the lands left free when the glaciers withdrew. Many forms are extensively cultivated. Hazel-nuts are used in the manufacture of confectionery.

English oak

Quercus robur, syn. *Q. pedunculata*, Fagaceae

This majestic tree is common all over Europe. Its Latin name is derived from the Celtic *caer quer*, meaning beautiful tree. The Greeks considered the oak to be sacred to Zeus and the Romans to Jupiter. At Dodona in Epirus, the site of the famous oracle of Zeus, prophecies were made by listening to the whispering of the wind in the leaves of the oak tree.

English oak

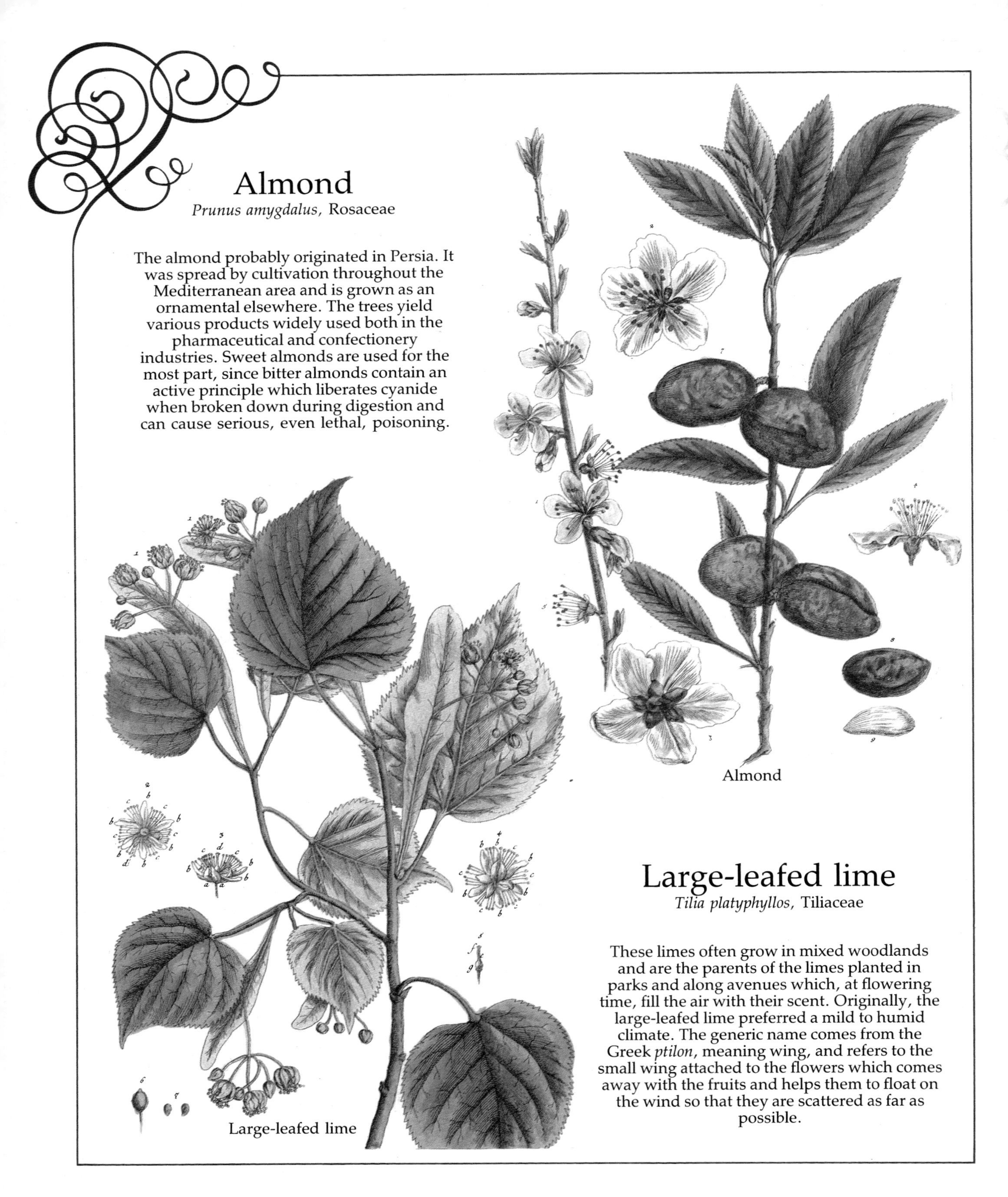

Almond

Prunus amygdalus, Rosaceae

The almond probably originated in Persia. It was spread by cultivation throughout the Mediterranean area and is grown as an ornamental elsewhere. The trees yield various products widely used both in the pharmaceutical and confectionery industries. Sweet almonds are used for the most part, since bitter almonds contain an active principle which liberates cyanide when broken down during digestion and can cause serious, even lethal, poisoning.

Almond

Large-leafed lime

Tilia platyphyllos, Tiliaceae

These limes often grow in mixed woodlands and are the parents of the limes planted in parks and along avenues which, at flowering time, fill the air with their scent. Originally, the large-leafed lime preferred a mild to humid climate. The generic name comes from the Greek *ptilon*, meaning wing, and refers to the small wing attached to the flowers which comes away with the fruits and helps them to float on the wind so that they are scattered as far as possible.

Large-leafed lime

Hops

Humulus lupulus, Cannabaceae

The generic name is thought to be derived from the Latin *humus,* meaning earth, because the stems trail on the ground if not supported. Young shoots can be eaten in spring like asparagus but the plant is mainly cultivated for its female flowers which grow in cones and are used to flavour beer.

Hops

Alder buckthorn

Alder buckthorn

Rhamnus frangula, Rhamnaceae

The alder buckthorn and other species of the genus have been known since antiquity mainly because of the purgative properties of the bark and fruits. The buckthorns are also a source of dyes which, before the discovery of aniline dyes, were much used both in artists' colours and in the textile industry.

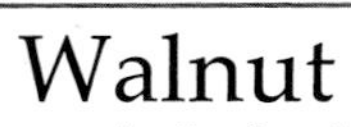

Walnut

Juglans regia, Juglandaceae

Native to Asia Minor and cultivated throughout Europe since ancient times, the walnut was sacred to Jupiter: *juglans* is short for *Jovis glans*, Jupiter's acorn.

Walnut

The kitchen garden and the wood

The interest in agriculture

There was great enthusiasm for the countryside during the 18th century. Profound transformations had taken place. The 'three field' system of crop rotation had begun to change in western Europe. Whereas cereals, legumes and fallow land had been rotated each year, fallow land was now planted for cattle fodder. Profits and accumulated capital from commercial activities were increasingly invested in land, both for personal pleasure and for economic security. From his country seat, a gentleman could 'watch cows walking in the sweet clover, their udders swelling, until the farm girl, watchful of the hour, milks them, singing in the morning sunshine'. In Holland, at the end of the 17th century, a prominent noblewoman from Amsterdam took up the cultivation of pineapples and an East India Company official grew the first coffee plant and the first banana tree in a greenhouse.

The importance of land as the basis of the economy was recognized by the physiocrats, a group of political economists who held that 'the natural order' governed society. As Lemercier de la Rivière wrote in 1777, 'all our interests, all our activities and intentions eventually merge to form, for our common happiness, a harmony which can be regarded as the gift of a benevolent divinity'. According to the physiocrats, agriculture was the only activity resulting in a 'net product', when a farmer obtained more wheat from his field than was necessary for his livelihood and for re-sowing. Hence they considered farmers the 'productive' class, whereas merchants, industrialists, doctors and lawyers belonged to the 'sterile' class. This theory superseded the mercantile system, according to which the accumulation of precious metals with monetary value was the basis of a nation's wealth.

The 18th century was also a practical age. Count Gasparo Gozzi wrote in 1782: 'Having a small kitchen garden at my disposal, I am reading books on agriculture and, among others, the works of Columella. To tell you the truth, the more I read him the more I love him. How many things were first said by him, only then to bring glory to others! And how fine and humorous his style is! and how full of flavour and satirical truth! But, in the end, what am I going to do with all this reading? I shall plant, following all the rules, God knows how many rows of beet and carrots, and I shall write down my experiences for posterity; because after all, as Columella says, experience is the real mistress of agriculture . . . not theory or science'.

The English philosopher John Locke had already expressed these sentiments, when he wrote that anyone wishing to own beautiful gardens and fertile fields should consult the slow-witted but experienced farmhand and the illiterate gardener rather than the profound philosopher or acute polemicist.

One of Locke's contemporaries, Peter Lauremberg from Rostock, wrote a *Horticultura* based on his own experiments. One of these involved planting 200 species of vine together with two varieties of cabbage to demonstrate the fallacy of an ancient belief, propounded in classical texts, that the two plants were harmful to each other. Another experiment with peas, marrows, walnuts, almonds and dates demonstrated, to his great satisfaction, that whatever the position of the seeds when planted, the seedlings always grew upwards and the roots downwards. This invalidated another inveterate belief according to which the position of the seed in the ground was important to the healthy growth of the plant.

Rousseau as botanist

In 1765 Rousseau withdrew to St Peter's Island in the middle of the Lake of Bienne, near Neuchâtel. He had been stoned by the inhabitants of the nearby town of Môtiers in the Val de Travers, where he had sought refuge from the authorities after publication of his *Social Contract* and *Emile*. The island offered him temporary peace. 'In place of sad paper-heaps and all that book trade, I filled my room with flowers and seeds; because I was then in the first fervour of my botanising, for which the doctor of Invernois has inspired in me a taste which soon became a passion. Since I did not wish to work any more at writing, there was necessary for me an amusement which pleased me, and which gave me no more trouble than that which a lazy man cares to give. I undertook to make the Flora of St Peter's Island, and to describe all the plants there, without omitting one, in sufficient detail to occupy me for the rest of my days. They say that a German has written a book on lemon-peel; I would have done one on each grain of the fields, on every moss of the wood, on each lichen which carpets the rocks; finally I did not want to leave a blade of grass, a vegetable atom which was not fully described.

'In consequence of this fine project, every morning after breakfast, which we took together, I went, a magnifying glass in hand, and my "System of Nature" under my arm, to visit a portion of the island, which I had for this purpose divided into small squares, with the intention of going over them one after the other, in each season. Nothing is more singular than the ravishments, the ecstasies which I felt at each observation I had made upon the structure and the vegetable organisation, and upon the play of the sexual parts in the fructification, of which the system was then altogether new to me. The distinction of generic characters, of which I had not beforehand the least notion, enchanted me . . .

'At the end of two or three hours, I returned laden with an ample harvest, a provision of amusement for the afternoon at home, in case of rain . . . while the others were still at the table, I escaped, and threw myself alone into a boat which I rowed into the midst of the lake, when the water was calm: and there, stretching myself out at full length in the boat, my eyes turned towards heaven, I let myself go and wander about slowly at the will of the water, sometimes during many hours, plunged into a thousand confused but delicious reveries, which, without having any well-determined object, nor constancy, did not fail to be in my opinion a hundred times preferable to all that I have found sweetest in what are called the pleasures of life.' (*The Reveries of a Solitary*, translated by John Gould Fletcher, 1927.) After six weeks Rousseau left St Peter's Island.

The role of the microscope

Collecting species, botanizing, describing, classifying organisms became a universal passion. The plant world was being discovered. Taxonomists honed down their concepts; Linnaeus' system reigned supreme and Rousseau walked about with the Swede's *Systema naturae* under his arm. Others laid the foundations for 'natural methods' of classification, based on the natural affinities between plants.

One of the first people to use a microscope was Robert Hooke (1635–1703), Professor of Geometry at the University of Oxford. He first placed a sliver of cork under a microscope and observed that it was made up of pores, and then did the same with the pith of the elder. Hooke did not know that cork is a vegetal substance; he reached that conclusion on the basis of the similarities in structure between cork and plant matter. He called the pores 'cells' and the name stuck. As a mathematician and astronomer, Hooke was more interested in the movements of the planets than plants, and from this stemmed his interest in improving optical instruments. In 1665 he published his *Micrographia, or some physiological descriptions of minute bodies made by magnifying glasses*, in which the term 'cell' was first used. The work was in fact an exercise in experimental philosophy: he wanted to demonstrate that our senses cannot perceive everything and that reality is quite different under the microscope.

Six years later, in 1671, a paper by Nehemiah Grew and a book by Marcello Malpighi were both presented, by a curious coincidence, at a meeting of the Royal Society. Grew's paper was entitled *The Anatomy of Plants Begun*, Malpighi's book *Anatome Plantarum idea*. The two authors were both born in 1628, both had studied plants under the microscope and had made the same fundamental discoveries—probably the first great scientific discoveries about the plant world. Grew was appointed Curator of the Anatomy of Plants by the Royal Society in 1672, with a salary of £50 a year. In his definitive work, *The Anatomy of Plants* (1682), he acknowledged Malpighi's work, particularly his *Anatome Plantarum* of 1675.

Grew occupies an important place in botanical history, but Marcello Malpighi made discoveries in the field of human anatomy which are probably even more important than his contributions to botany. He graduated in medicine from Bologna and taught first at Pisa and later at Bologna. He is, however, more highly regarded today than he was in his own time when two anatomists publicly insulted him and tried to dissuade students from attending his dissections. Two other colleagues, aided by a band of thugs, attacked him and wrecked his house. However, the study of living matter had begun.

The popularity of oranges

Daniel Defoe spent most of his adventurous life writing polemical and sociological works. His *Tour through the Whole Island of Great Britain*, a guidebook published in 1724, contains a reference to oranges growing at

Beddington, Surrey. They had been cultivated there since the end of the 16th century and a hundred years later the trees were producing 10,000 fruits a year. 'The orange trees continue and are indeed wonderful; they are the only standard trees in England, and have moving houses to cover them in winter; they are loaded with fruit in summer, and the gardeners told us, they have stood in the ground where they now grow above eighty years.'

A 17th-century writer relates that ladies used to make bracelets and necklaces from young oranges as small as beads because of their sweet scent. This was just one of the customs which justified the cultivation of these plants, which was only possible in unsuitable climates if great care was exercised.

The word 'orange' is derived from the Persian *narang* which in turn comes from the Sanskrit *naranja*, meaning 'fruit favoured by the elephants'. Its etymology is an indication of how the orange spread to Europe from its native habitat. By orange, however, we mean the bitter orange, which Linnaeus named *Citrus aurantium*. Its country of origin was probably India. The Romans imported it from the East but may not have cultivated it. It spread to all the shores of the Mediterranean after the Arabs introduced it to Spain (from Egypt), hence its other name of Seville orange.

Tradition has it that the sweet orange, *Citrus sinensis*, now the most common citrus fruit in the world, was imported directly from China, where the Portuguese had long cultivated it. The name of João de Castro is mentioned in this respect. He was a naval commander who was appointed viceroy of Portuguese India in 1547 but who died a year later at Ormuz, having received the last rites from St Francis Xavier. However, the sweet orange may have in fact reached Europe earlier than this, since it is specifically mentioned in a document in the town archives of Fermo, in the Italian Marches, which dates from the 14th century.

Plant reproduction

Among the 34,000 volumes in Christoph Jakob Trew's library, there must have been a copy of *De sexu plantarum epistula*, a pamphlet published in 1694 by the Academy at Tübingen and written by a colleague of Trew's, Rudolph Jacob Camerarius. This little-known work was an account of experiments carried out to prove the sexual reproduction of plants. Camerarius' father was a Professor of Medicine at Tübingen where the German humanist Melanchthon had also taught in the 16th century, and he himself studied medicine there. He was nominated Extraordinary Professor of Medicine and Director of the botanic gardens. A year after the publication of his research, on the death of his father, he succeeded him as professor and spent the rest of his life in Tübingen.

The basis for Camerarius' experiments was the following set of general observations: in the majority of flowers the stamens are near the style; when mature, the anthers of the stamens split open and the stigma is dusted with pollen; some flowers have no stamens and a larger number of petals, in which case they do not produce seeds, as any gardener can verify. In the quiet of the Tübingen botanic garden, Camerarius postulated that the pollen was actually the 'male semen', and it then became obvious that many flowers were hermaphrodite.

The young scientist turned to the microscope, as the works of Grew and Malpighi had taught him to do. He was thus able to observe the formation of the plant embryo and to observe that it is only formed after the pollen has reached the pistil. He also noticed that some plants have hermaphrodite flowers, some bear both male and female flowers on the same individual (monoecious) and some bear the male and female flowers on separate individuals (dioecious). This division only applied to flowering plants, not to the whole of the vegetal kingdom.

Camerarius thought that in plants with hermaphrodite flowers, fertilization occurred within the flower itself (which is not quite true); while in monoecious and dioecious plants, an external agent—the wind which carried the pollen—would have to intervene. His experiments on dioecious plants, like the mulberry and spinach, and on monoecious species such as castor oil plants and maize, supported his theories. They involved preventing the male part from coming into contact with the female, to demonstrate that reproduction did not take place. Female mulberries grown without male plants in the vicinity bore fruits incapable of producing embryos. The seeds of female *Mercurialis* plants grown in isolation from male individuals were found to be sterile when sown. And when all the flowers with stamens were removed from castor oil plants, they produced only empty shells.

The pioneer of hybridization

The strawberry plant painted by Elizabeth Blackwell, which bears the generic name of *Fragaria*, was the wild strawberry. Almost all the large strawberries we eat today are hybrids of two American species, *Fragaria virginica* and *Fragaria chiloensis*. Their names reflect their different places of origin, one from Virginia and the other from Chile; they were crossed in France towards the end of the 18th century.

Plant hybridization was pioneered by a German botanist, Josef Gottlieb Koelreuter (1733–1806), who began by examining all the experiments on plant reproduction carried out since the time of Camerarius. Koelreuter was the son of an apothecary from Tübingen, was educated there, and later became custodian of the natural history collections in the Imperial Academy of St Petersburg. There he began his own experimental studies on pollination, which he completed in the Margrave's gardens in Karlsruhe, the capital of the state of Baden. In 1760, while in St Petersburg, he produced a hybrid between two species of *Nicotiana* (tobacco); his later experiments and his analyses of them show him to have been 'incontrovertibly one of the great forerunners of modern genetics and of scientific plant breeding' (A.G. Morton).

Koelreuter recognized another aspect of plant reproduction which is taken for granted today, but which at the time had barely been noticed and had never been observed scientifically: the role played by insects in pollination, and of nectar in attracting insects to the flowers. Koelreuter wrote that he was astonished to discover that plant fertilization could be left to chance, to a lucky accident. However, his astonishment soon changed to admiration for the method—at first sight accidental but in fact infallible—'which God had chosen to ensure the reproduction of plant species'.

A green valley. Daniel Defoe, *The Life and Surprising Adventures of Robinson Crusoe of York, Mariner*, 1718: 'I came to an opening, where the country seemed to descend to the west; and a little spring of fresh water, which issued out of the side of the hill by me, ran the other way, that is, due east; and the country appeared so fresh, so green, so flourishing, everything being in a constant verdure or flourish of spring, that it looked like a planted garden . . . I saw here abundance of cocoa-trees, and orange and lemon and citrus trees; but all wild and very few bearing any fruit; at least not then. However, the green limes that I gathered were not only pleasant to eat, but very wholesome . . .'

The Savoy Alps. Laurence Sterne, *A Sentimental Journey through France and Italy*, 1768: 'Nature! in the midst of thy disorders, thou art still friendly to the scantiness thou hast created—with all thy great works about thee, little hast thou left to give either to the scythe or to the sickle—but to that little thou grantest safety and protection; and sweet are the dwellings which stand so sheltered. Let the way-worn traveller vent his complaints upon the sudden turns and dangers of your roads—your rocks—your precipices—the difficulties of getting up—the horrors of getting down—mountains impracticable—and cataracts, which roll down great stones from their summits and block his road up.'

Memoirs of the Count O—. Friedrich Schiller, *The Ghost-seer* (1786–1789). 'It was my custom to travel on foot in fine weather, being more agreeable to me, and affording a better opportunity of observing the surrounding objects. I pursued it now. The tears rolled from my cheeks, when I arrived at the foot of a mountain. Above my head the branches of the trees formed a grove, through which we scarcely could perceive the valley underneath, which was ornamented by an opposite hill; upon which, oak trees, the produce of centuries, raised their majestic heads. I stood before a deep dale, and enjoyed the romantic prospect which presented itself to me. I was lost in the contemplation of it, and on each twig my eye dwelt with a voluptuous pleasure.'

A hurricane on the Ile de France (Mauritius). Jacques Henri Bernardin de Saint-Pierre, *Paul and Virginia*, 1787: 'One of the most aged of these planters, approaching the governor, said to him,—"We have heard all night hollow noises in the mountain; in the woods, the leaves of the trees are shaken, although there is no wind; the sea-birds seek refuge upon the land: it is certain that all these signs announce a hurricane." . . . Everything, indeed, presaged the near approach of the hurricane. The centre of the clouds in the zenith was of dismal black, while their skirts were tinged with a copper-coloured hue. The air resounded with the cries of tropic-birds, petrels, frigate-birds, and innumerable other sea-fowls which, notwithstanding the obscurity of the atmosphere, were seen coming from every point of the horizon to seek for shelter in the island.

'Towards nine in the morning we heard in the direction of the ocean the most terrific noise, like the sound of thunder mingled with that of torrents rushing down the steeps of lofty mountains. A general cry was heard of, "There is the hurricane!"—and the next moment a frightful gust of wind dispelled the fog which covered the Isle of Amber and its channel . . . Every billow which broke upon the coast advanced roaring to the bottom of the bay, throwing up heaps of shingle to the distance of fifty feet upon the land; then, rushing back, laid bare its sandy bed, from which it rolled immense stones, with a hoarse and dismal noise. The sea, swelled by the violence of the wind, rose higher every moment; and the whole channel between this island and the Isle of Amber was soon one vast sheet of white foam, full of yawning pits of black and deep billows. Heaps of this foam, more than six feet high, were piled up at the bottom of the bay; and the winds which swept its surface carried masses of it over the steep sea-bank, scattering it upon the land to the distance of half a league. These innumerable white flakes, driven horizontally even to the very foot of the mountains, looked like snow issuing from the bosom of the ocean. The appearance of the horizon portended a lasting tempest: the sky and the water seemed blended together. Thick masses of clouds, of a frightful form, swept across the zenith with the swiftness of birds, while others appeared motionless as rocks. Not a single spot of blue sky could be discerned in the whole firmament; and a pale yellow gleam only lightened up all the objects of the earth, the sea, and the skies.'

A forerunner of Darwin

Georges Louis Leclerc de Buffon was born in 1707, the same year as Linnaeus, whom he survived by ten years. They were both contemporaries of Dr Trew, Mrs Blackwell and Ehret. In 1739 Buffon was appointed Curator of the Jardin du Roi, the greatest centre for the study of natural history then in existence. Three years later Linnaeus became Professor at Uppsala. They spent the rest of their lives, in their respective positions, immersed in prodigious endeavours of great importance not only to specialists but to all men of learning.

The Comte de Buffon was the son of a Dijon lawyer, born in the family castle of Montbard, in Burgundy. He travelled with the Duke of Kingston to the south of France, Italy and Switzerland, and then spent a few months in London. He became known as a result of scientific pamphlets and experiments in physics, one of which confirmed Archimedes' theory of burning-glasses. Having been appointed Curator of the royal gardens, he began his great work, *Histoire Naturelle* (Natural History). The first three volumes, *Théorie de la Terre et vues générales sur la génération et sur l'homme* (Theory of the Earth and general views on creation and on mankind) were published in 1749; they were followed by twelve volumes on quadrupeds (1753–67), nine on birds (1770–83), five on minerals (1783–88) and seven volumes of supplements: thirty-six volumes in all of an immense work of literary as well as scientific value, to which a few collaborators contributed.

The lord of Montbard only spent four months a year in Paris, dealing with the royal gardens and putting in an appearance at a few of the famous salons of the 'philosophes'. He spent the rest of the year on his estate. His study was a room in an old tower on a hill overlooking the grounds. He would go there every morning at five (his secretary would be there waiting for him), and he would work, read, study and dictate until two in the afternoon. It is said that when writing he would wear lace cuffs—an indication of his character, perhaps.

It was difficult to reconcile his *Théorie de la Terre* with Genesis, and many believers were offended, although Buffon himself was a Christian. In 1751 the Sorbonne condemned fourteen of the book's propositions. The preface to the fourth volume was a letter in which Buffon declared he had abandoned everything which might be contrary to Moses' teachings; and the foreword to one of the supplements, *Les Epoques de la Nature*, published in 1778, stated that nothing in it disagreed with the Scriptures, provided that the Bible was not interpreted literally whenever it appeared to contradict reason and the facts of nature. A quarter of a century had gone by since the first volume of the *Histoire Naturelle* had appeared, the cultural climate had changed, the author had become an important figure and a member of the French Academy; the Faculty of Theology, alarmed once more, was asked by the King to reserve criticism, and complied.

Whereas Linnaeus, in his *Philosophia botanica*, had written that one could count as many species as there were different forms created in the beginning, Buffon propounded his theory of the 'degeneration of the original species', which places him among the precursors of the theory of evolution. 'One might as well say,' he hypothesized, 'that the monkey belongs to the same family as man, that it is a degenerate man, or that the monkey and man have a common origin.' The seeds of Darwinism had been sown.

The stimulating effects of coffee

The quotation from Voltaire which opens this chapter mentions coffee from Batavia and coffee from 'the American islands' (that is, the French West Indies), together with Mocha. As such it contains the broad outlines of the plant's history. According to an Arab legend, the coffee plant and the custom of drinking coffee were discovered by a sheik who fled to the mountains of the Yemen for religious reasons. The Ethiopians probably introduced coffee to the Yemen during one of their invasions, since the plant is certainly a native of tropical Africa. At the end of the 17th century, the Dutch took it from Arabia to Batavia, the capital of Java, in the course of their trade in the Indian Ocean. From there, coffee beans were introduced into European botanical gardens and the French took them to their colonies in the West Indies at the beginning of the 18th century. Curiously enough, when French agronomists first experimented with cultivating the plant on the island of Réunion, they discovered a very similar native species.

Eighteenth-century Europe was swept by a craze for coffee. According to the Milanese writer and economist Count Pietro Verri, who wrote a *Storia naturale del caffè* (Natural history of coffee), it arrived in Marseilles in 1644 and the first coffee shop in Europe opened there in 1671. A century or so later a Parisian wrote that the consumption of coffee in France had trebled, it was offered in all the houses of the bourgeoisie, every shopkeeper, cook and maidservant breakfasted on *café au lait*, and the porters in the Halles 'drink coffee standing up, their baskets on their backs'.

Verri sang the praises of the heartening and stimulating effect of coffee: 'It awakens the mind, in many it keeps sleep at bay, and is particularly useful to people who take little exercise and cultivate scientific pursuits.' Indeed, because of the 'restorative virtues' of the plant, he named the periodical he published *Il Caffè*. The group of liberal intellectuals who gathered around him, and his brother, Alessandro Verri, were contributors.

Il Caffè was the most influential periodical of the Italian Enlightenment, an expression, like Dr Trew's publication of the botanical drawings of Blackwell and Ehret, of the Age of Reason. The spirit of curiosity towards the world and towards nature which resulted from a rejection of superstition, tradition and religious dogma in favour of a belief in man's power as a rational being, inspired the developments touched on in this book—the discovery of new worlds, and the advent of botany as a science.

Christoph Jakob Trew, a Nuremberg physician

1695
C.J. Trew was born into a solid but not particularly wealthy middle-class family. His father was an apothecary at Lauf an der Peignitz, a village a few kilometres from Nuremberg. His grandfather Abdias was Professor of Mathematics, Physics and Astronomy at Altdorf University.

1711–1717
He studied medicine at Altdorf, where a place had been reserved for him from the age of twelve. He gained his doctorate thanks to the financial help of a cousin; this allowed him to continue his studies which had been briefly interrupted after graduating.

1717-1720
The travelling years. Thanks to a scholarship granted by the city of Altdorf, Trew set off on a journey which was to complete his education. He visited Frankfurt, Strasbourg, Basle, Berne, Zurich, Lausanne, Geneva, Lyon and finally Paris where he spent 13 months. France, and Paris in particular, was considered the homeland of science. While in Paris, a lucky encounter with two noblemen from Altdorf allowed Trew to continue his travels; he accompanied them to Amsterdam via Brussels, Louvain and Antwerp. In Amsterdam he met a man from Danzig, and, all expenses paid, accompanied him to northern Germany—to Bremen, Hamburg, Lubeck and to Danzig where he spent two months.

1721
On his return to Lauf, Trew applied unsuccessfully for the Chair in Surgery and Anatomy at Altdorf University, but was accepted as 'ordinary physician' by the Collegium Medicum, the society of Nuremberg doctors. He moved to Nuremberg and spent the rest of his life there, practising medicine. One of his duties was to direct the Collegium's botanical garden, and he only relinquished the directorship in 1731, in favour of a younger colleague. He wrote in his letters: 'Botany is one of the most pleasurable occupations . . . so appealing that once sampled one cannot give it up'. And years later: 'I recommended the study of botany, not only because it is seemly, worthy of note and useful to a doctor, but also because, since early times, there has always been a member of the Nuremberg Collegium who excelled in it, while now I am the only one who has modestly undertaken it'.

1725
A successful doctor and lecturer in anatomy at the Collegium Medicum's anatomy theatre, Trew also gave lectures, botanical demonstrations and organized teaching expeditions in his capacity as Curator of the botanic gardens. He began to collect books and specimens of natural history. In 1725 F.E. Brückmann wrote in his *Epistulae intinerariae:* 'In the house of C.J. Trew I saw an anatomical museum, collections of shells, minerals, stones and ores, marine plants, exotic trees, and a splendid library containing works on natural history'.

1730
Trew formed a society with other physicians to publish a weekly magazine on scientific subjects, the *Commercium literarium ad rei medicae et scientiae naturalis incrementum institutum* (Literary exchange founded for the amelioration of medicine and the natural sciences). An annual subscription cost between two and three florins, depending on the quality of the paper. That year the archaeologist J.G. Keisler, who had visited Nuremberg, wrote: 'In the anatomical theatre instituted by the town one can see one hundred skeletons of animals, particularly birds. From the ceiling hang the intestines of a man, to prove that human intestines are six times as long as a man's height. Doctor Trew is the director of this theatre in which well devised inscriptions can be seen.

1732
Having relinquished his post as Curator of the botanical garden, Trew planted his own garden to grow rare plants. He planted 150 in 1732, including thirty-two varieties of aloe.

1733
Georg Dionysius Ehret visited Trew in Nuremberg. The artist had been introduced to him by J.A. Beurer, the town's assistant apothecary who was then studying in Regensburg. Trew commissioned Ehret to draw botanical illustrations at one florin each. Thus began a long association. Trew later wrote: 'After we became friends, dear Mr Ehret enriched and adorned my naturalistic collections for over thirty years. Of the illustrations he drew for me, the best are those of rare, exotic plants grown in England, which he painted with great aptitude and the art which comes naturally to him'.

1734
Trew's medical practice was flourishing; five years after he set it up, 'it had grown so much that I could hardly wish for more'. Having been a physician for fourteen years, Trew estimated his income at 800–900 thalers a year.

1735
Trew became director of the *Commercium literarium*, the scientific magazine he had founded five years previously. He summarized its aim thus: to establish a wide-ranging correspondence throughout Europe in the fields of physics and medicine, and to make known to the public the remarkable achievements taking place in these sciences. Trew wrote one hundred or so articles for the magazine, but the running of it became an increasing burden. Publication ceased in 1745, after fifteen years.

1736
The Margrave of Brandenburg-Ansbach appointed Trew his personal physician and court counsellor. Trew successfully treated several members of the Margrave's family, but he never agreed to live at court as the prince wished.

1744
Trew was appointed Director of the Imperial Academy of Naturalists, as well as Imperial Counsellor and Count Palatine of the Holy Roman Empire. He served as director of the Academy for twenty-five years, until his death. Among those he sponsored for membership were Ehret and Albrecht von Haller, the scientist whose visit from Casanova we have related. In 1744 Trew also bought from various owners the unpublished 16th-century manuscripts of the Swiss Konrad Gesner. After pressure from several quarters to publish this vast collection of botanical illustrations, he entrusted the work to Casimir Schmiedel, of Erlangen University. The first volume appeared between 1751 and 1753; the second in 1771, after Trew's death; a third was planned but never published.

1750
The first ten folios of *Plantae selectae*, illustrated by Ehret, were published. Printed on German paper, they cost 3.43 florins, and on Dutch paper, 3.38 florins. The work was to be completed in five years, folios appearing every six months, but it soon became delayed; three more folio collections appeared in 1754, a fifth in 1756, the sixth and seventh in 1768; the last three were published in 1775, after Trew's death, by B.C. Vogel, Professor of Medicine and Botany at Altdorf. The publication of the first ten folios was greeted with enthusiasm. Albrecht von Haller wrote: 'Since the year 1750 Trew has published a splendid work in ten-folio volumes . . . Flora has nothing more marvellous than this.' And Linnaeus, who had dedicated a species to Trew (*Trewia nudiflora*) wrote: 'Miracles of our own century in natural science . . . such as antiquity never saw and posterity hardly will.' In 1750 the first volume of the *Herbarium Blackwellianum* was also published: this was a redrawn and corrected edition of Elizabeth Blackwell's *Curious Herball*. Christoph Jakob Trew did not live to see this project completed, as was the case with Gesner's drawings and the *Plantae selectae*.

1761
At the age of sixty-six Trew was appointed *senior primarius* of the Nuremberg Collegium Medicum: this was the highest honour in the medical profession.

1769
The year of his death. The Margrave of Brandenburg-Ansbach bestowed upon him the title of Privy Councillor. The decree arrived just after the scientist had died.

Bibliography

Alpino, Prospero, *Giornali Veneziani del Settecento*, ed. Marino Berengo, Milan, 1962.

Arber, Agnes, *Herbals, their origin and evolution*, London, 1953.

Bertoldi, Vittorio, *Parole e Idee*, Paris, 1927

Blunt, Wilfrid, *In for a Penny. A prospect of Kew Gardens*, London, 1978.

Blunt, Wilfrid and Raphael, Sandra, *The Illustrated Herbal*, London, 1979.

Bonora, Ettore, ed. 'Letterati, memorialisti e viaggiatori del settecento' in *La letteratura italiana*, Milan, 1951.

Carletti, Francesco, *Giro del mondo del buon negriero*, ed. E. Radius, Milan, 1941.

Caygill, Marjorie, *The Story of the British Museum*, London, 1981.

Charliat, Pierre-Jacques, 'Le temps des grands voiliers' in *Histoire Universelle des Explorations*, ed. L. H. Parias, 1960.

Chiej, Roberto, *Medicinal Plants*, London, 1984.

Colombo, Fernando D., *Le Histoire della vita e dei fatti di Cristoforo Colombo*, Milan, 1930.

Duby, Georges and Mandrou, Robert, *Storia della civiltà francese*, Milan, 1968.

Goldoni, Carlo, *Memoirs*, trs. John Black, London, 1814.

Gualino, Lorenzo, *Saggi di Medicina storica*, Turin, 1930.

Heilmann, Peter, preface to *Das Kräuterbuch der Elizabeth Blackwell*, Dortmund, 1981.

Launert, Edmund, preface to *Erlesene Pflanzen* by Christoph Jakob Trew, Dortmund, 1981.

Margotta, Roberto, *Medicina nei secoli*, Milan, 1967.

Masefield, G. B., Wallis, M., Harrison, S. G., Nicholson, B. E., *The Oxford Book of Food Plants*, Oxford, 1969.

Morton, A. G., *History of Botanical Science*, London, 1981.

Pirson, Julius, 'Der Nürnberger Arzt und Naturforscher Christoph Jakob Trew' in *Festschrift des Vereins für Geschichte der Stadt Nürnberg*, Nuremberg, 1953.

Stuart, David G., *The Kitchen Garden*, London, 1984.

Tonzig, Sergio, in *Enciclopedia della Scienza e della Tecnica*, Milan, 1963.

Verri, Pietro, *Il Caffè*, ed. Sergio Romagnoli, Milan, 1960.

Elizabeth Blackwell's herbal included this illustration of *Corallum rubrum*. Coral was used for medicinal purposes, and apothecaries' shops in England used to stock an *electuarium diacorallion*. Linnaeus classified corals and madrepores as the sixty-fifth plant order, but they were deleted by the contemporary French botanist, Bernard de Jussieu, who proved that they were animals.

Index of Plants

For Latin names see illustration plates.

Absinthe, 101
Abutilon, 83
Acacia, 89
Adonis, Spring, 34
Agrimony, Hemp, 101
Almond, 118
Aloe, 43, 64, 130
Angelica, 46, 65
Apple
  Custard, 103
  May, 84
  Thorn, 33, 67
Arctotis, 76
Arnica, 50
Artichoke, 101–2
  Jerusalem, 97
Asparagus, 111
Aubergine, 97
Azalea, 15
Azarole, 79

Balm, Lemon, 65
Banana, 78, 95, 96–7, 99
Banyan tree, 104
Begonia, 101
Belladonna, 49, 58, 63
Birch, Silver, 116
Bird-of-paradise flower, 101
Birthwort, 40
Black-eyed Susan, 100
Bocconia, 79
Boehmeria, 18
Bonus Henricus, 54
Breadfruit tree, 95
Bromelia, 83
Buckeye, Red, 82
Buckthorn, Alder, 119
Bugloss, 66
Burning bush, 44
Butterbur, 51

Cacao, 88, 94
Calamus, 64, 65, 66
Camellia, 100
Cardoon, 101
Cassava, 85, 95
Ceanothus, 71
Cecily, Sweet, 55
Cedar of Lebanon, 13
Celandine, Greater, 101
Cereus, 77
Cherimoya, 92, 103
Chervil, 55
Chicory, Wild, 55
Chinese lantern, 112
Cinchona, 59
Clove, 32, 58, 65
Coca, 94
Cochlearia, 55
Cocoa, 98–9
Coconut palm, 95
Coffee, 58, 98, 99, 111, 128
Coltsfoot, 47
Columbine, 52
Coral tree, 86
Coriander, 65
Cornflower, Blue, 40, 101
Crataegus, 79
Creeping Jennie, 101
Crocus, Saffron, 57, 66
Crotalaria, 76
Cuckoo-pint, 38
Cyperus, 66

Dahlia, 100
Dandelion, 48
Datura, 66, 67
Deadnettle, Large red, 18

Elder, 45
Elecampane, 41

Fennel, 110
Fenugreek, 54
Fern, Maidenhair, 63
Fig, 19, 84, 100
Frangipani, 91
Fuchsia, 100

Gardenia, 58
Gelatophyllis, 66
Gentian, 64
Ginseng, 74
Grape vine, 110
Guava, 86

Hart's-tongue, 42
Hawthorn, 79
Hazel, 116
Helenium, Egyptian, 66, 67
Hellebore, 59, 60
Hemp, Indian, 58, 67
Henbane, 38, 66, 67
Herb Robert, 101
Hestiateris, 66
Holly, Sea, 49
Hollyhock, 30
Honeysuckle, 31
Hops, 119
Horse-chestnut, Common, 82
Hyacinth, 55
Hymenocallis, 81
Indigo, 74
Iris, 16
  Japanese, 104
Ivy, 50
Ixia, 80

Jimson weed, 67
Juniper, 100

Knapweed, 40

Larch, European, 114
Ledum, 12
Lily, 17, 54
  Belladonna, 80
  Blood, 88
Lily-of-the-valley, 34
Lime, Large-leafed, 118
Liquorice, 39
Liverwort, 62
Lotus, 66–7
Lovage, 42
Lungwort, 14
Lychee, 103
Lycium, 70

Magnolia, 73
Maize, 94, 95, 96
Mallow, 54
  Dwarf, 30
  Marsh, 36, 54
Mandrake, 36, 56, 58, 59, 66, 67
Manihot, 85
Manioc, 95
Maple, 78
Mezereon, 32
Milfoil, 101
Monarda, 70
Money-wort, 101
Moth mullein, 16

Narcissus, 55, 101

Oak, English, 117
Oenothera, 66
Orange, 113, 123–4

Papaya, 87, 95
Passion flower, 97
Peony, 45, 54, 59, 62
Pepper, 97
  Red, 44
Peppermint, 46
Periploca, 20
Petiveria, 82
Phyllanthus, 85
Pine, Stone, 115
Pineapple, 69, 94, 95, 96
Plantain, Ribwort, 11
Pomegranate, 12
Pontederia, 71
Poplar, Balsam, 75
Poppy, Opium, 29, 58
Potato, 94, 96
  Sweet, 95

Rhododendron, 15
Rhubarb, 64, 97
Robinia, 100
Rose, 31, 54
  Christmas, 37, 59
Rosemary, 39

St John's wort, 62, 101
Sandbox tree, 90
Sassafras, 20
Sedge, Sweet, 64
Sensitive plant, 75
Sida, 89
Soursop, 90
Spruce, Norway, 114
Squill, Sea, 48
Strawberry, 62–3, 109, 125
Sumach, 97
Sweetsop, 90

Tea, 98, 99
Tephrosia, 14
Terebinth, 66
Thistle
  Carline, 100
  Stemless, 35
Tobacco, 93–4, 125
Tomato, 97, 112
Tulip, 55
Tulip tree, 72

Valerian, 52, 58, 101

Walnut, 63, 120
Watercress, 55
Willow, Virginia, 72
Wormwood, 101

Yam, 95
Yarrow, 101
Yucca, 85, 97

Zamia, 92

Index of names

Adanson, Michel, 106
Aiton, William, 24, 25
Albertus Magnus, 27
Alione, Dr, 23
Alpino, Prospero, 21, 60, 102–3
Amadio, Andrea, 23
Anderson, William, 108
Arber, Agnes, 94, 102, 104
Arnaldo di Villanova, 66
Auenbrugger, Leopold, 68
Augusta, Princess of Saxe-Gotha, 24, 25, 53

Badonia, Juan, 94
Banks, Sir Joseph, 24, 26, 101, 107, 108
Baré, Jeanne, 107
Baretti, Giuseppe, 23, 26
Bégon, Michel, 101
Benincasa, Giuseppe (Goodenhuyse), 102–3, 104
Blackwell, Alexander, 8–9, 54
Blackwell, Elizabeth, 8, 9, 10, 22, 26, 53, 54, 125, 127, 128, 130
Bligh, Captain William, 95, 108
Blunt, Wilfrid, 23
Bocconi, Paolo, 79
Boehmer, G. R., 18
Boerhaave, Herman, 8, 54, 60–61, 67, 99
Bonafede, Francesco, 21
Bougainville, Louis Antoine de, 107
Bromel, Olof, 83
Brown, John, 58
Brown, Lancelot "Capability", 25
Brunfels, Otto, 101
Buffon, Georges Louis Leclerc, Comte de, 127–8
Busbecq, Augier Ghislain de, 55, 56, 104
Bute, Earl of, 25, 26, 100

Camellus, George Joseph, 100
Camerarius, Rudolph Jacob, 124–5
Carletti, Francesco, 96, 98, 103–4
Caroline, Queen, 24, 25, 26
Carrara, Francesco da, the Younger, 23
Carteret, Captain, 106–7
Casacarius, Pastor, 104
Casanova, Jacques, 61–2, 130
Castro, João de, 124
Catherine de' Medici, Queen, 93
Caygill, Marjorie, 54
Celsus, Aulus Cornelius, 59–60
Chambers, Sir William, 25
Charlemagne, Emperor, 35, 39, 100
Charles XI, King of Sweden, 99
Charles de l'Ecluse *see* Clusius
Charlotte Elisabeth of Bavaria, Princess Palatine, 99
Chasseboeuf, Constantin François de, Count of Volney, 65
Chinchon, Countess of, 58
Chomel, Pierre Jean-Baptiste, 60
Cibo, Gherardo, 22
Clifford, George, 10, 99
Clusius, 27, 55, 103, 104
Colonna, Fabio, 101
Columbus, Christopher, 58, 93, 94, 95, 96
Columbus, Ferdinand, 93, 94
Columella, L. Junius Moderatus, 121
Commerson, Philibert, 107
Cook, Captain James, 24, 25, 96, 107–8
Cosimo I, 21, 22
Culpeper, Nicholas, 42
Cuvier, Georges, 106

Dahl, Anders, 100
Dampier, William, 101, 105, 106
De la Cruz, Martin, 94
Defoe, Daniel, 123, 126
Della Porta, Giambattista, 62, 63
Diderot, Denis, 107
Dioscorides, Pedanios, 22, 42, 54, 55, 56, 58, 59, 70
Dodoens, Rembert (Dodonaeus), 27
Drake, Sir Francis, 104
Drakenstein, Van Rheede tot, 104
Duck, Stephen, 25
Durante, Castore, 26, 101

Ehret, Georg Dionysius, 9–10, 22, 24, 26, 96, 100, 103, 104, 108, 127, 128, 130
Eisenberger, N.F., 9
Evelyn, John, 24

Fabricius, Johann Albert, 66
Filippo, Jacopo, 23
Forskål, Peter, 65
Forster, Johann Georg, 108
Forster, Johann Rheinhold, 108
Frederick, Prince of Wales, 24, 25, 53
Frederick V, King of Denmark, 65
Froben, Johann (Frobenius), 63
Fuchs, Leonhard, 56, 97, 100, 104
Füllmaurer, Heinrich, 56

Galen, Claudius, 42, 48, 57, 63
Gaza, Theodore, 56, 100
George II, King, 53, 54
George III, King, 24, 25, 26
Gerard, John, 26, 27
Gesner, Konrad, 8, 130
Ghini, Luca, 21–2, 23, 56, 61
Goethe, Johann von, 21
Goldoni, Carlo, 67, 68
Gozzi, Count Gasparo, 28, 121
Grassi, Giovanni Battista, 59
Grew, Nehemiah, 123, 124
Gualino, Lorenzo, 67

Haid, Johann Jakob, 7, 10, 24
Haller, Albrecht von, 61–2, 130
Hamon (Suleiman's physician), 57
Hernandez, Francisco, 94
Herodotus, 67
Hippocrates, 48, 57
Hohenheim, Theophrastus Bombastus von *see* Paracelsus
Holmes, Oliver Wendell, 58
Hooke, Robert, 123

Isidore of Seville, 100

Jenner, Edward, 68
Jussieu, Bernard de, 10
Jussieu, Joseph de, 59, 106

Kaempfer, Englbrecht, 53, 104–5
Kalm, Peter, 106
Kamel, George Joseph, 105
Karl III, Margrave of Baden-Durlach, 9
Keisler, J.G., 129
Koelreuter, Josef Gottlieb, 125
Krateuas (Greek herbalist), 55, 56, 66, 101

La Condamine, Charles Marie de, 59, 106
Lauremberg, Peter, 122
Liberale of Udine, Giorgio, 57
Linnaeus (Carl von Linné), 9, 10, 18, 24, 25, 30, 34, 36, 58–9, 61, 65, 71, 73, 83, 86, 89, 90, 98–102 *passim*, 106–9 *passim*, 123, 124, 127, 128, 130
Lobelius (Matthias de Lobel), 27, 102
Locke, John, 121
Loeschenkohl (banker), 9–10, 104
Ludwig, Christian Gottlieb, 9

Magnol, Pierre, 73
Malpighi, Marcello, 100, 123, 124
Marsili, Dr, 23
Matteo di San Giuseppe, Fr, 104
Mattioli, Pierandrea, 20, 22, 30, 56–7, 61, 70, 109

Mead, Richard, 10
Merini, Fr Michele, 22
Mesmer, Franz, 68
Meyer, Albert, 56
Meyerpeck, Wolfgang, 57
Miller, Philip, 10
Mithridates VI Eupator, King of Pontus, 55, 66, 101
Monardes, Nicolò, 70
Moraeus, Sara, 102
Morgagni, Giovanni Battista, 68
Morton, A.G., 27, 54, 100, 125
Müller, Gabriel, 7
Musa, Antonio, 96

Nearchus, 104
Nelson, David, 108
Nicot, Jean, lord of Villemain, 93
Niebuhr, Carsten, 65

Orta, Garcia da, 67

Paracelsus, 62, 63
Parkinson, John, 27
Pericoli, Niccolò (Il Tribolo), 22
Petiver, James, 82
Philip II, King of Spain, 94
Pinell, Philippe, 60
Plantin, Christophe, 27, 28
Pliny, 8, 12, 17, 54, 60, 66, 67, 96, 109
Plumier, Charles, 91
Plutarch, 65–6
Poinsett, Joel, Roberts, 101
Poivre, 32
Pontedera, Giulio, 71
Portland, Duchess of, 10

Ray, John, 53, 105
Richer (physicist), 106
Rinio, Benedetto, 23
Rivière, Lemercier de la, 121
Robin, Jean, 100
Rothman, Dr Johann, 99
Rousseau, Jean-Jacques, 62, 122, 123
Rudbeck, Olof, 99
Ruel, Jean, 101

Saint-Pierre, Jacques Henri Bernardin de, 126–7
Schiller, Friedrich, 126
Schmiedel, Casimir, 130
Serapion the Younger, 23, 67
Sherard, William, 105
Sibthorp, Humphrey, 10
Silvius
see Bocconi, Paolo
Sloane, Sir Hans, 8, 10, 24, 53–4, 57, 99, 105
Solander, Dr Daniel Charles, 107, 108
Sparrman (Swedish doctor), 108
Sterne, Laurence, 54–5, 126
Sydenham, Thomas, 57, 58

Talbot, Sir Robert, 58
Theophrastus (Tirtanus), 13, 16, 30, 45, 54, 56, 60, 89, 100
Thunberg, Carl Peter, 100
Tonzig, Sergio, 100
Tournefort, Joseph Pitton de, 36, 53, 88, 96, 99
Trew, Dr Christoph Jakob, Count Palatine of the Holy Roman Empire, 7–10 *passim*, 24, 26, 61, 68, 88, 96, 103, 104, 108, 124, 127, 128, 129–30
Tribolo, Il
see Pericoli, Niccolò

Valdez, Oviedo y, 69
Vasco da Gama, 87
Verri, Count Alessandro, 60, 61, 128
Verri, Count Pietro, 128
Vogel, B.C., 130
Voltaire, François Marie Arouet de, 62, 107, 109, 128

Wallis, Samuel, 106
Weinmann, Johann Wilhelm, 9